Photovoltaic Power Generation

Solar Energy R&D in the European Community

Series C

Photovoltaic Power Generation

Volume 4

Solar Energy R&D
in the European Community

Series C Volume 4

Photovoltaic Power Generation

Proceedings of the
EC Contractors' Meeting held in
Hamburg/Pellworm, 12-13 July 1983

edited by

W. PALZ
Commission of the European Communities

D. Reidel Publishing Company

A MEMBER OF THE KLUWER ACADEMIC PUBLISHERS GROUP

Dordrecht / Boston / Lancaster

for the Commission of the European Communities

Library of Congress Cataloging in Publication Data
Main entry under title:

Photovoltaic power generation.

(Solar energy R&D in the European community. Series C, Photovoltaic power generation: v. 4)
1. Photovoltaic power generation-Congresses.
I. Palz, W. (Wolfgang), 1937- . II. Commission of the European Communities. III Series.
TK2960.P465 1984 621.31'244 83-26960
ISBN-13: 978-94-009-6344-3 e-ISBN-13: 978-94-009-6342-9
DOI: 10.1007/978-94-009-6342-9

Organization of the Contractors meeting by
Commission of the European Communities
Directorate-General Science, Research and Development, Brussels

Publication arrangements by
Commission of the European Communities
Directorate-General Information Market and Innovation, Luxembourg

EUR 8830

Softcover reprint of the hardcover 1st edition 1984

Published by D. Reidel Publishing Company
P.O. Box 17, 3300 AA Dordrecht, Holland

Sold and distributed in the U.S.A. and Canada
by Kluwer Academic Publishers,
190 Old Derby Street, Hingham, MA 02043, U.S.A.

In all other countries, sold and distributed
by Kluwer Academic Publishers Group,
P.O. Box 322, 3300 AH Dordrecht, Holland

INTRODUCTION

This book provides an overview of the results achieved during the European Community's photovoltaic pilot programme which was terminated on 30 June 1983.

Reports contained herein concentrate on the construction phase of the pilot projects. It should be noted that the results of the design phase were published before as Volume 1 of this same Series C of the Reidel Company.

The programme aim was the indigenous European development of a comprehensive set of photovoltaic systems and of all associated technologies. The power range from 30 to 300 kW was covered, including the world's largest fixed-plate array to date.

The programme was initiated by the Commission in 1980 within the European Community's second Energy R+D Programme. For its implementation the Commission's services in Brussels assumed general management responsibility. While in the design phase, coordination meetings of project leaders were organized at the Commission's offices in Brussels, those meetings were held later on at various construction sites, namely in Crete (GR), near Mont Bouquet (F) and finally near Pellworm (D). The proceedings of this latter meeting are the main subject of the present book.

A particular feature was that it was possible to pool together 15 different plants which are spread all over Europe and involve many different technological approaches in one single programme. Even better, the programme has become a good example for the effectiveness of a European and trans-national approach for large and complex research projects in this area.

The pv pilot plants were developed and built by consortia or industrial associations from 8 Member Countries of the European Communities. It was a remarkable and significant outcome that mutual understanding and cooperation have developed among the major pv companies in Europe in the course of the programme.

After completion of most of the plants today, it can be said that the programme was successful. In particular the following goals set out by the Commission in the beginning could be met :

- a large variety of high performance pv technologies and systems are now available in Europe ready to meet many applications;
- photovoltaic systems are viable in all European climates;
- photovoltaic systems can be built environmentally and aesthetically attractive;

Europe's industry now has a firm base of valuable technology to build on for future photovoltaic applications in and outside Europe.

W. Palz

C O N T E N T S

REVIEW PAPERS

Overview of the European Community's Photovoltaic Pilot Programme

W. PALZ
Programme Leader

Commission of the European Communities

At the beginning of 1980, the Commission launched a call for tenders for photovoltaic pilot projects in the 30 to 300 kW range. After a selection procedure 15 projects were started at the beginning of 1981.

Table 1 gives a list of the 11 projects which were completed on schedule by the end of June 1983.

The first plant to enter operation in October 1982 was the one in Crete. Experience gained with this first plant was excellent: the plant worked very reliably and delivered as much as 15000 kWh of electricity to the village making any back-up by Diesel sets unnecessary.

Table 2 shows the 4 projects to be completed later on, i.e. in June 1984 at the latest.

It is expected that the late project in England will be complete in October 1983 and the three projects in Italy in the first semester of 1984.

Total budget of all 15 projects was approximately 30 million ECU. One-third of this cost was borne by the Commission in the frame of the European Community's 2nd Energy R+D Programme (1979-83). The rest of the budget was contributed by the Member Countries and other institutions as indicated on Tables 1 and 2. It is important to note that all projects were also heavily co-funded by the industry itself; in some cases as high as 50%.

The Commission's services in Brussels assumed general management responsibility throughout the programme.

Modules. The modules employed in this programme are representative of the state-of-the-art in the field. Modules were supplied by AEG and Siemens from Germany, France Photon and Photowatt from France, Pragma and Ansaldo from Italy, BP Solar from England and IDE/Fabricable from Belgium. Modules were qualified by the Commission's Joint Research Centre at Ispra in a special test campaign.

Single and polycrystalline silicon cells are employed. Module size ranges from 20 to over 100 W and involves various encapsulation techniques. In general, European module development has benefited considerably from this pilot programme.

Array. The internal array loss due to mismatch of modules, cabling and diodes, a measure of the quality of the array, could generally be kept as low as 2%.

Power conditioning. All plants use AC systems except three, one of which is linked to a water electrolysis unit, another to a water disinfection unit and the other to DC compression motors. Inverters are solid state with one exception (motor). Inverters are either designed for grid connection or stand-alone operation, some are for both. Some systems use maximum power point trackers, others do not. The importance of managing power flow to ensure safe, reliable and efficient operation has been recognized and microprocessor control has been widely used to this end. High efficiency of approximately 90% even at 10% load or 10% of max. possible insolation was achieved either in simple systems with one inverter or multiple switching units.

A full range of DC/AC inverters from 1 kVA to 75 kVA has been developed by nine different European manufacturers: Jeumont Schneider, Aérospatiale and Leroy Somer, France, ETCA, Belgium, Filectron and Italenergie, Italy, AEG and Siemens, Germany, Holec, Netherlands.

Support structures. Most developments had to start from scratch. All structures are fixed, some allowing seasonal adjustment. Two arrays are roof-integrated; the others are installed either on flat roofs or on the ground. Design criteria depend on the wind load and hence from the site. Weight per m^2 could be strongly reduced in comparison with typical structures employed hitherto. Many different designs and construction materials (metal or wood) were used. Costs depend very much on the terrain but in several cases structure cost did not exceed 3 to 4% of the total plant cost, an important achievement.

Batteries. They were improved designs provided by Varta, Germany or Oldham, UK.

Monitoring. Plants are generally unattended as they need no maintenance. All plants emoloy the same standard monitoring system. Raw data are transmitted for exploitation to Ispra acting as the Commission's monitoring centre for this programme.

For plant acceptance, the Commission's Joint Research Centre, Ispra, has developed a new device which measures the power of sub-arrays up to 10 kW. 11 plants were accepted this way by the end of July.

TABLE 1

EEC pv pilot plants

completed June 30, 1983

Site	Power	Application	Leading contractor	Co-sponsor
Pellworm Island (D)	300 kW	Power for recreation centre	AEG	German Government
Kythnos Island (GR)	100 kW	Island grid	Siemens	German Govt., Greek utility PPC
Chevetogne (B)	63 kW	Swimming pool, lighting	IDE/ACEC	Belgian Government
Aghia Roumeli (GR)	50 kW	Village grid	Renault	French Govt., Greek utility PPC
Mont Bouquet (F)	50 kW	TV emitter	Photowatt	French Govt., Télédiff. de France
Nice Airport (F)	50 kW	Tower control	Photowatt	French Govt., Chamber of Commerce
Fota Island (IRL)	50 kW	Dairy farm	U. Cork, AEG	Irish + German Govts., Irish utility
Terschelling (NL)	50 kW	Marine school	Holec	Dutch and German Govts.
Kaw (French Guy.)	35 kW	Village grid	Renault	French Government
Hoboken (B)	30 kW	Electrolysis	ENI	Belgian Government
Rondulinu, Corsica (F)	30 kW	Village power	Leroy-Somer	French Govt. and French utility EDF

TABLE 2

EEC pv plants to come

Site	Power	Application	Leading Contractor	Co-sponsor	Completion date
Marchwood (UK)	30 kW	Grid interaction	BP Solar	UK Government	October 1983
Tremiti Island (I)	65 kW	Seawater desalination	Italenergie	Italian Government	March 1984
Giglio Island (I)	45 kW	Water disinfection, cold store	Pragma	Italian Government, Toscana, Giglio Community	June 1984
Vulcano Island (I)	80 kW	Village power	ENEL	Italian utility ENEL	June 1984

STATUS OF U.S. LARGE PV SYSTEMS APPLICATIONS

Dr. H. L. Macomber
Gaithersburg, Maryland, U.S.A.

SUMMARY

This report presents a summary of the Intermediate size (15 kW - 1000 kW) photovoltaic system applications which have been installed or are being planned for installation in the United States of America. The systems summarized include those funded under the U.S. National Photovoltaic Program and those funded and constructed by the U.S. private sector. The systems are described and operating characteristics of installed systems summarized. Design and operating problems are presented to provide information for future designers and manufacturers.

1. INTRODUCTION

Since 1977 the U.S. National Photovoltaic Program has actively supported the design, construction and operation of a variety of photovoltaic power system applications classed as Intermediate size systems and ranging in size from 15 kW to 1000 kW. To date fourteen systems have been built and generally are operating as planned. One 100kW application project was terminated before the completion of the project due to exhaustion of funds and too ambitious of a project. The fourteen operating systems represented a total installed peak generating capacity of 1,164 kW and have delivered nearly 2,000 MWH of energy.

Under the U.S. National Program two additional intermediate size photovoltaic power plants are under construction and are expected to become operational during 1984. These two plants represent a total of 1,300 kW of peak generating capacity and should deliver over 1800 MWH of energy per year.

In addition the U.S. Photovoltaic Program has committed most of the funding to support the design, procurement and construction of a 1000 kW photovoltaic plant. This plant will be constructed and installed during the 1984-1985 period. Current U.S. Program Plans do not anticipate funding any further Intermediate size photovoltaic power plants.

In parallel with the government program, the private sector has begun building and installing Intermediate (and larger) size photovoltaic power systems. Over the past 21 months, the private sector has installed four Intermediate size PV systems in the U.S. These four systems represent an installed peak generating capacity of 1,286 kW. It is estimated that these systems should be delivering energy of more than 2,000 MWH per year.

The private sector systems have been subsidized in whole or in part by a variety of tax benefits available under U.S. Federal and some State tax laws. The use of these tax codes to attract private investors to fund the installation and operation of Intermediate (and larger) size photovoltaic systems has encouraged the U.S. photovoltaic industry to seek such project funding. In addition, U.S Federal laws requires U.S. electric utilities (private or public) to interconnect small (less than 80 MW), renewable energy generators to their system and to pay a fair price, based upon an "avoided cost" scheme, for any energy produced by the renewable energy system and fed into the utility system grid.

It is estimated that private sector financing of Intermediate size PV plants using tax benefits for investors could result in an additional 20 to 30 MW of installed power generating capacity before the end of 1985. One supplier has announced plans for installing a 16 MW PV power plant by 1985 with 6 MW operating by the spring of 1984. Contracts for selling the power to the utility and engineering the system have been signed and the site has been selected.

This paper summarizes the technical features of the U.S. Intermediate size PV plants and highlights some performance and design features of the operating systems to date.

2. U.S. GOVERNMENT PROGRAM SYSTEMS

Table I lists the seventeen installed, planned or cancelled systems funded by the U.S. National Photovoltaic Program. The first four systems listed in Table I have been turned over by the U.S. Department of Energy (D.O.E.) to the application site operator for maintenance and operations. As a consequence no operating data has been collected or processed by the D.O.E. since the end of 1981. Although these four earlier systems have experienced problems, which will be summarized below, they do continue to operate and provide useful power to the application.

Each of the systems listed on Table I will be described in the following paragraphs. Details of these systems may be found in a variety of D.O.E. and conference publications including (1). Information on the current status is contained in (2).

Mead, Nebraska Agriculture Application

In July of 1977, the MIT Lincoln Laboratory and the Agricultural Engineering Department, IANR, University of Nebraska, Lincoln, installed a 25 kW PV array at the UNL Field Laboratory at Mead, Nebraska. This was the first large PV array and consists of Solarex and Sensor Technology modules. The array has been utilized for research programs which have included the development of an automated irrigation system for 80 acres of corn, a natural-air grain drying system and an electric-arc process for producing nitrogen fertilizer, the latter in cooperation with Kettering Foundation Laboratories.

Since 1977, data has been collected on the completely instrumented solar system. A 20 hp DC motor and 10, 15 and 20 hp AC motors were used to drive the surface irrigation system. Relative efficiencies, battery storage requirements, and operational characteristics were evaluated. Grain drying and nitrogen generator experiments were also conducted. Important operational concerns such as load matching, battery storage, power conditioning, array dirt accumulation, module failures and safety were investigated.

During the first two years of operation module failure rate was only 2 percent of the total number of modules in the array. Since that period failure rate has increased but the weak design features of the modules causing the failures have been corrected in later models.

System reliability, however, has continued to be excellent despite individual module failures.

Mt. Laguna, California Radar Station

This system was developed under a cooperative program between the U.S. Department of Defense and the D.O.E. It consists of a 60 kWp flat plate array interconnected with a Helionetics/DECC DC to AC inverter to the stations diesel generator powered grid. The PV system represents about 10 percent of the total station power generating capacity.

While there have been early and continuing module failures in the array due largely to "hot spot" problems of these early module designs, system performance has been good. Overall array degradation has been severe due to the module problems. Improved module designs for later

TABLE I U.S. GOVERNMENT PROGRAM
INTERMEDIATE SIZE PV APPLICATIONS

APPLICATION	LOCATION	SIZE (kWp)	CONSTRUCTION COMPLETE	FULL OPERATIONS	AVAILABILITY %	TOTAL ENERGY TO MARCH '83 (MWH)
AGRICULTURE	MEAD, NE	25	JULY 1977	JULY 1977	>90	NO DATA PROC. AFTER 11/81
RADAR TRACKING STATION	MT. LAGUNA, CA	60	AUG. 1979	AUG. 1979	NOT AVAIL	NO DATA COLL. AFTER 12/81
AM RADIO STATION (WBNO)	BRYAN, OH	15	AUG. 1979	AUG. 1979	NOT AVAIL	NO DATA COLL. AFTER 11/81
HOUSING/VISITOR CENTER NATIONAL PARK	NATURAL BRIDGES NAT'L MONUMENT, UT	100	JUNE 1980	JUNE 1980	NOT AVAIL	DATA BEING PROCESSED
UPS FOR NEWMAN SUBSTATION (UTILITY)	EL PASO, TX	20	DEC. 1980	JAN. 1981	96	63.6 (DC)
HIGH SCHOOL	BEVERLY, MA	100	FEB. 1981	JUNE 1981	85	161.4 (AC)
SHOPPING CENTER	LOVINGTON, NM	100	MAR. 1981	MAR. 1981	99	382.0 (DC)
COMMUNITY COLLEGE	BLYTHEVILL, AR	240 ACTIVE (P. TROUGH CONC.)	MAY 1981	OCT. 1981	93 (MANNED)	33.5 (AC)
LIGHT MFG. FACILITY	SAN BERNARDINO, CA	35	SEPT 1981	-	80	24.4 (AC)
HOSPITAL	KAUAI, HI	35 ACTIVE (P. TROUGH CONC.)	SEPT 1981	-	95	3.7 (AC)

TABLE I U.S. GOVERNMENT PROGRAM
INTERMEDIATE SIZE PV APPLICATIONS (CON'T)

APPLICATION	LOCATION	SIZE (kWp)	CONSTRUCTION COMPLETE	FULL OPERATIONS	AVAILABILITY %	TOTAL ENERGY TO MARCH '83 (MWH)
PUBLIC SCIENCE AND ARTS BUILDING	OKLAHOMA CITY, OK	135 (FP/MIRROR)	FEB. 1982	DEC. 1982	65	53.7 (AC)
SKY HARBOR AIRPORT	PHOENIX, AZ	225 PASSIVE (FRESNEL CONC.)	APR. 1982	DEC. 1982	95	206.0 (AC)
AIRPORT UTILITY PLANT LIGHTS/BOILER	DALLAS-FT. WORTH AIRPORT, TX	27 ACTIVE (LINEAR FRESNEL CONC.)	JUNE 1982	SEPT 1982	98	12.1 (DC)
OFFICE BUILDING BDM CORPORATION	ALBUQUERQUE, NM	47 ACTIVE (P. TROUGH CONC.)	JUNE 1982	DEC. 1982	20	6.7 (DC)
JUNIOR COLLEGE	SENATOBIA MS	101	STOPPED	-	-	-
GEORGETOWN UNIVERSITY INTERCULTURAL ACTIVITIES BUILDING	WASHINGTON, D.C.	300	IN PROCESS SCHEDULED FOR 1984	-	-	-
CENTRAL UTILITY PHASED DEVELOPMENT (UTILITY AND CALIF. STATE COST-SHARING)	SACRAMENTO MUNICIPAL UTILITY DISTRICT (SMUD), SACRAMENTO, CA	1000 (PHASE 1)	IN PROCESS SCHEDULED FOR 1984	-	-	-

TABLE II PRIVATELY BUILT INTERMEDIATE SIZE PV SYSTEM EXAMPLES

BUILDER/SUPPLIER	APPLICATION	LOCATION	SIZE (kWp)	DESIGN FEATURES	CONSTRUCTION COMPLETE
SOLAREX CORPORATION	PV FACTORY	FREDERICK, MARYLAND	200	STAND-ALONE, INTEGRATED PV/BLDG/PROD. EQUIP.	OCT. 1982
COPLEY ENERGY INC./ SOLAREX CORPORATION	CENTRAL UTILITY DEMONSTRATION, DEL MARVA POWER & LIGHT CO.	DENTON, MARYLAND	15.7	LOW-COST FIXED STRUCT.	DEC. 1982
EXXON/SOLAR POWER CORPORATION	EPCOT CENTER - PAVILION/UTILITY (REEDY CREEK UTILITY)	WALT DISNEY WORLD ORLANDO, FLORIDA	70	ROOF MOUNTED, BIDIRECTIONAL ENERGY FLOW	1982
ARCO SOLAR/ BDM CORPORATION	CENTRAL UTILITY, SOUTHERN CALIFORNIA EDISON CO., LUGO SUBSTATION	HESPERIA, CALIFORNIA	1,000	2-AXIS F. P. TRACKERS	DEC. 1982
ARCO SOLAR/ FLUOR CORPORATION	CENTRAL UTILITY, PACIFIC GAS & ELECTRIC CO.	CARRISA PLAIN, SAN LUIS OBISPO COUNTY, CALIFORNIA	16,000	2-Axis F. P. TRACKERS, REFLECTOR-ENHANCED	FIRST 6MW SCHEDULED FOR MARCH 1984

modules have accounted for "Mt. Laguna" type failures. Thus, this system has been an excellent test bed for developing large system array and overall power plant designs.

Bryan, Ohio Radio Station WBNO

In August 1979 radio station WBNO in Bryan, Ohio became the first radio station in the U.S. to be powered by a PV system. The system has a 15 kW flat plate array mounted on a simple, bolted galvanized steel rack designed to minimize site preparation and foundation expense.

System loads are DC thereby eliminating the need for an inverter. Minimal battery storage has been included in the system.

The array on this system has suffered similar failures of the early module designs as the prior applications. Contact interconnect failures of the modules have continued due to the earlier design. All failed modules have been replaced with later design modules by the manufacturer under the array warranty and the system continues to operate.

Natural Bridges National Monument, Utah

This 100 kWp system became operational in June 1980. It provides all of the electrical power required for the monument workers residences and the visitor building. The system incorporates a DC - AC inverter and 750 kWh of lead-acid battery storage. A 40 kW diesel generator provides emergency back-up to the PV system.

Module shorting problems were experienced and have been overcome. Two batteries (1% of the total number of battery cells) have failed. The system otherwise has continued to operate as designed.

El Paso, Texas, Newman Utility Substation

System startup for this 20 kWp system occurred on January 27, 1981. The power generated displaces rectified AC power used by generator control equipment tied to an uniterruptible power supply (UPS). The operating voltage of the UPS, 134 volts DC, controls the voltage of the ungrounded PV system. the PV array has 64 panels of 9 series-connected modules. The 64 panels are connected in parallel and are tied to the UPS through a DC switchboard that controls the daily sequencing of the system.

The system has operated successfully for the past two years, producing more than 60 megawatt hours of energy. At the conclusion of this contract, the system will be operated and maintained by the site owner, El Paso Electric Company.

The success of this project has demonstrated that intermediate-sized PV systems are technically feasible. They can be built and operated by personnel with no prior PV experience. They can be designed for unattended operation, and high reliability of prime system components can be expected.

Beverly, Massachusetts High School

This 100 kWp system provides electrical power to the Beverly High School in Beverly, Massachusetts. The array is a fixed-tilt, ground mounted structure located on a small hill by the school. Construction was completed in February 1981 and the system became fully automatic in June 1981.

The two DECC inverters (Helionetics, Inc.) which control the system operation are designed to shut down automatically upon detection of any one of twelve different fault conditions. In each case the inverter will remain shut down until manually reset. During the first 10 months of operation sporadic ground faults were the main cause of fault shutdowns. After reducing the ground fault sensitivity in mid January 1982, these sporadic ground faults were no longer a problem. But during January and February the system was plagued by shutdowns due to synchronization errors. These were caused by control circuits going out of range in cold weather. The problem was cured by replacing the temperature-sensitive components on the PC boards in early March. The system was essentially free of fault shutdowns during March and April, but a major malfunction of the computerized data acquisition system (ODAS) caused a series of ground fault shutdowns in May. These ended when the ODAS was disconnected and sent out for repair. In June, one inverter began to blow fuses due to an intermittant problem in one or two of the inverter legs, resulting in one half of the system being down for most of the month. This problem also occurred once in July. A number of shutdowns due to excessive grid voltage also occurred throughout one week in July. These ended when the utility fixed the voltage regulator at its substation. A few shutdowns due to overtemperature also occured in July because the ventilation sytem in the inverter house had not been working properly. August was relatively free of fault shutdowns, but shutdowns of one inverter due to logic power errors began to occur frequently in September. A dirty pot in the bias power supply which feeds the control circuitry was causing the problem, which disappeared after the pot was cleaned in early October.

Both inverters were given a complete checkout in early October. Since then, fault shutdowns have been very rare and have not caused any significant downtime up to the present. For the full year of 1982, total system output was 78% of the predicted value. The shortfall can be accounted for entirely by the downtime caused by the various fault shutdown problems described. Based on the essentially fault-free operation over the past eight months, these problems now appear to be resolved.

Lovington, New Mexico, Shopping Center

This system is a 100 kWp utility grid-connected system. This flat plate system has been operating automatically and unattended since March 1981. It has produced more than 375 MWH of energy during its first two years of operation. It provides about 8 percent of the annual energy requirements of the Lovington Square Shopping Center.

The system consists of two equal subfields, each providing power to and controlled by a 60 kW rated power conditioning unit. Each unit provides peak power tracking and inverts the nominal 240 VDC array output to 480V, 3 phase, 60 Hertz AC and controls the power transfer to the grid.

The PV array structure of galvanized steel frames may be adjusted manually to 10°, 30° or 40° from the horizontal. Experiments have been conducted with each subfield array tilted at different angles. For example, during the period from August 1981 through March 1982 one subfield was tilted at 30° and the other at 40° from the horizontal. The subfield tilted 40° produced 3,600 kWh more than the 30° tilted subfield.

The Lea County Electric Cooperative is operating and maintaining the system, with the New Mexico Solar Energy Institute providing some technical support. Data analysis will continue through the Phase III contract.

Reliability of prime system equipment has been high, with few actual hardware failures. However, frequent "resets" of the power conditioning units and the data acquisition system have been necessary. A daily site visit is made to check system status. Actual maintenance time, including site visits, has averaged 1.7 hours per megawatt hour of production.

Blytheville, Arkansas, Mississippi County Community College

This project consists of a uniquely energy-conserving building with a floor area of 5100 square meters and 63,000 PV solar cells mounted on 270 single-axis tracking, parabolic trough concentrating collectors. The local condition rating of the system is 240 kWp electrical and it is designed to provide about 85 percent of the total energy needs of the building. Thermal energy for heating the building is obtained from the ethylene glycol fluid used to actively cool the solar cells.

The system contains a DECC DC-AC inverter and two desk top computers for controlling the system and data acquisition. It is a utility interactive system where excess electrical energy is fed back to the local utility, the Arkansas Power and Light Company, on a one for one exchange ratio with backup power fed by the utility.

The system performance has never reached its full capability. Drooping reflectors, diffuse tracking and receiver glass rippling have contributed to the low performance. In addition the system has suffered from severe ground fault problems. Nevertheless, the system does continue to operate and provide useful energy to the school.

San Bernardino, California Light Manufacturing Facility

This 35 kWp flat plate array system has been operational since September 1981. The array structure is mounted on a stand-off structure attached to the flat roof of the light manufacturing facility to which it provides power. The system is utility-interactive and provides only a portion of the facilities electrical needs.

Module failure problems on this array have been due to some design and quality control problems. The system also has been plagued with moisture-induced ground faults. As a result system performance has never reached design capacity.

Kauai, Hawaii, G. N. Wilcox Hospital

The Wilcox Hospital photovoltaic (PV) project system is rated at 35 kW of direct current electrical power and 200 kW (0.68 million Btu/hr) of thermal energy at the design condition of 1,000 W/m^2 of direct normal insolation and 50°C cell temperature.

Funding was provided in part (77 percent) by the U.S. Department of Energy (DOE) under its Concentrating Photovoltaics Application Experiment Program, with the State of Hawaii providing funding (23 percent) administered through the Hawaii Natural Energy Institute (HNEI).

The Wilcox Hospital in Lihue, Kauai was chosen as the site for the PV system because the hospital can use both the electrical and thermal energy provided by the PV system. The main energy loads for the hospital are air-conditioning and hot-water heating. The air-conditioning requirement peaks near midday and in the summer, corresponding to solar energy availability. The Hawaiian location at 22°N latitude provides excellent insolation, although the Lihue climate proved to be less than ideal for a concentrating solar system.

Electricity is generated from PV cells along the focal line of the parabolic trough concentrators. The PV collector field is made up of 10 groups consisting of 8 collectors in each row. Each collector is 6 ft. by 10 ft. in aperture. The collectors are oriented with their axes aligned along the true north-south direction for minimum incidence and end losses. Each receiver contains 3 strings of 36 cells in parallel. A bypass diode is in parallel with each group of three cells to maximize fault and shadowing tolerance. The resulting electrical power is converted and conditioned by the power conditioning system to supply alternating current electricity at 490V, 3 phase to the hospital, eliminating the necessity for electrical storage.

The thermal energy and cell cooling system consists of a closed-loop coolant circuit; a 3,000-gal storage tank with an immersion-type heat exchanger for heating the hospital potable water; and a coolant circulation pump and finned tube cooling convectors for rejecting excess heat. The coolant fluid is water that has been treated to boiler feedwater standards.

The power conditioning unit (PCU) has two distinct functions. The first is to hold the field at its peak power point, and the second is to produce alternating current power compatible with the utility. The peak power tracker and inverter are controlled by a microprocessor system incorporated into the unit. The output of the inverter can be fed directly into the main electrical distribution bus at the hospital in parallel with the utility service.

Two major problem areas were encountered during system startup. The first was degradation of the cell modules on the photovoltaic receivers, which had not previously occurred during hundreds of hours of prototype testing. The PVB material which bonds the cells to their glass superstrate was bubbling. This overheating of the PVB was caused by a manufacturing quality control problem and was corrected by opening the receivers and adding sufficient thermal grease.

The second problem area was the apparent inability of the PCU to sustain operation in parallel with the utility grid. This problem was caused by voltage fluctuations in the Kauai electric grid. Modifications were made to the unit which restored reliable operation.

The poor weather during 1982, the worst in 50 years, prevented the system from producing the expected level of energy. In addition, due to the extended weather-related downtime, moisture penetration in the receiver resulted in fungus growth and corrosion of PV cell interconncects. The fungus growth is inhibited when the receiver operates at high temperature.

The weather has improved significantly in 1983. The PV receivers appear to have reached a steady-state condition and no major degradation has been noticed over the last year of operation.

Despite the initial problems, the system is currently operating reliably. After 1.5 years of operation, the system is performing at 70 percent of the design prediction. The system is in a fully automatic condition and does not require any onsite operator; although grid transients cause occasional problems and trip the PCU.

Oklahoma City, Oklahoma Center for Science & Arts

This system is a 135 kW utility interactive photovoltaic power system located at the Oklahoma Center for Science & Arts, Oklahoma City, Oklahoma.

The PV array is based upon the polycrystalline silicon technology of Solarex Corporation. System modules are composed of a 6 x 12 cell arrangement with complete series/parallel interconnection to provide optimal power capacity and reliability in combination with the enhancement mirrors. The array is tilted at 30° with the corresponding mirrors angled to the north at 39°. The mirrors provide reflective enhancement from March through October with peak output occurring in late June.

Electrically two groups of 9 strings, containing 84, 5 volt, 14 amp, modules each, compose the array. By design, at test conditions the array operates at 420 volts, 252 amps and is capable of delivering 105 kW without enhancement and 140 kW with peak enhancement.

The power conditioning system is composed of a Windworks, Inc. 150 kW Gemini, 3 phase, 6 pulse line commutated static power conditioner. The unit operates from 320 VDC to 450 VDC input with a 360 VAC phase to phase output. Isolation and step up to the utility 480 VAC four wire termination are performed with a 180 KVA static drive transformer.

Currently the system is fully operatonal after experiencing some early break-in failures. Two major operational anomalies surfaced in the first year:

1. A 15% difference in array peak capacity based upon sample string measurements and factory simulator measurement

2. Power conditioner induced electrical interference with facility sound and lighting equipment

Interference from the system power conditioner with high performance audio and special effects lighting has occasionally forced temporary shutdowns of the entire power sytem. Investigation and preliminary tests indicate the problem is potentially attributable to the presence of the PCUs commutation notch on the facility electrical system. Power system equipment failures have been chiefly attributable to equipment breakin, and required the system to be shutdown for much of the first few months of operation. The data acquisition system has also accumulated a track record for failures, maintenance and downtime, which has made evaluation of the systems performance innovations difficult. Operation and maintenance has been contained to occasional equipment inspections and instrumentation observation. Due to local climate conditions and nominal air pollution levels, array cleaning will only be necessary once annually.

The system observation/test activities have continued through the spring and early summer 1983. During this period further investigation is being made into the effects of mirror enhancement, maximum power tracking and the cause of the reduced array capacity. Several equipment implementation/revision tasks directed at improving the data acquisiiton system and eliminating the power conditioner induced interference problems are being performed. Additionally, consideration is being given to the implementation of an intelligent system monitor/controller, which will provide a power conditioning system nighttime shutdown feature and several other life cycle cost saving functions.

Phoenix, Arizona Sky Harbor Airport

This project, one of four concentrator system application experiments sponsored by D.O.E. first delivered solar generated electricity in April, 1982. The peak kW array field is comprised of 80 Martin Marietta "first generation" passively cooled tracking arrays, which feed a Power Systems and Controls inverter. The AC power output is stepped up to 12.6 kV and fed directly into the Arizona Public Service (APS) power grid at the Airport site. Each concentrator uses a Frensnel lens to focus energy (36X) on a 2.25 inch diameter silicon solar cell. A total of 272 lens-cell combinations are mounted onto a pedestal mounted, two-axis tracking array structure. The system was constructed at a cost of approximately four million dollars.

The initial operating experience has been generally favorable, although several problem areas have been identified. The first year's net AC energy production was about 2000,000 kWh, about half the amount expected from predictive models. The most frequently occurring operational problems have been in the array tracker drive, inverter control, current-voltage testing, and module assembly subsystems. During the second operating year, the full-time two man operating crew will be removed from the site to see how the system behaves on its own.

Dallas-Fort Worth, Texas Airport

The Dallas-Fort Worth (DFW) Airport photovoltaic and thermal (PVT) application experiment has been operational since June 21, 1982. the system was formally dedicated in July, checkout-tested in August, and has performed on a seven-day-per-week basis since 1 September 1982. The collector field consists of 110 Linear Fresnel lens collector modules mounted in eleven independent tracking groups, called ten-collector arrays. The total collector field aperture area is 245 square meters. The arrays are mounted on the roof of the Central Utility Plant at the airport. The collector field's electrical and thermal outputs are used to power lights and to preheat boiler feedwater, respectively, within the plant.

Analysis of system performance over the first six months of operation provides the following results: [i] the PCU has had a large amount of down-time, with a six-month availability of only 84%; [ii] the monthly electrical efficiency is strongly dependent on the monthly PCU availability; [iii] the monthly thermal and total efficiencies are relatively constant, despite widely varying levels of insolation, fluid temperature, and ambient temperature over the six-month period; and most importantly, [iv] the bottom-line efficiencies for the six-month period are 6.8% electrical, 38.5% thermal, and 45.2% total.

To determine if the system is achieving expected performance levels, the system simulation computer program has been modified to predict what the system electrical and thermal performance should be under actual measured operating conditions. The total (electrical plus thermal) output shows excellent analytical/empirical agreement. However, the electrical output alone is slightly lower than predicted levels in the morning, but close to predicted levels in the afternoon. This aberration is due to an antenna which shades the collector field during the morning, degrading the electrical output of some of the arrays. This antenna is currently being moved. Furthermore, performance has been affected by jet fuel soiling of the collectors. System electrical performance should improve significantly when the antenna move is completed.

<u>Albuquerque, New Mexico - BDM Office Building</u>

The system was dedicated in July of 1982. This is a nominal 50 kilowatt peak system located on the roof of the BDM facilities in Albuquerque, New Mexico. The photovoltaic concentrating array is a single-axis linear parabolic trough utilizing single crystalline silicon photovoltaic cells under 41.56 suns of net solar concentration. A total of 7,560 square feet of aperture is provided by 54 Solar Kinetics T-700 7-foot aperture mirrors configured in nine north-south oriented rows. A two-piece 90° V-shape receiver places the cells at the focal point of the mirror and provides cooling utilizing an ethelyne glycol/water mixture. The photovoltaic array is designed to operate in parallel with the local utility in an augmentary load sharing mode. A portion of the thermal energy resulting from the cell cooling is used for building heating during the winter months.

The major subsystems are the photovoltaic array, DC and AC power conditioning, thermal system, and the control and data acquisition systems. The array is composed of the concentrating mirror, receiver tube and the PV-cell and module. The power conditioning includes the Westinghouse inverter and the wiring and switchgear. The thermal system involves both a heat rejection circuit and a building heating circuit utilizing ground coupled heat pumps. Components of the control and data acquisition system are the collector control loops and the instrumentation transducers. The overall control and safety monitoring of the system is accomplished automatically by an AIM-65 micro-computer. The controls and instrumentation equipment are located in a central monitoring area found near the building entrance for easy visitor access. The basis of the data acquisition system is a government furnished HP-9845 computer driven acquisition system.

The system is currently operating fully automatically 7 days a week including holidays. Final alignment of the photovoltaic receivers was completed in late March 1983 and it is expected that the system should perform near its design point during summer solstice.

Initial operation of the array resulted in a slight discoloration and minor bubbling of the photovoltaic modules at design operating temperatures. This was observed during the first few weeks of operation and has not progressed since the original occurrence. Some module shorting to substrate has also occurred, but to date the use of spares and the repair of these modules has allowed a full complement of modules to remain in the field. It appears that ASEC did not properly isolate some of the internal wiring during the final assembly processes. Tracking with the Solar Kinetics shadow band trackers has been less than optimum and has been one of the major operational headaches. The mechanical portions of the system have operated well with only minor maintenance required. The entire mirror field was cleaned in January 1983 with a resulting substantial increase (20-30%) in power. Previous cleaning resulted in only 7-9% power increases; however, more attention will be paid to this area of operation and maintenance in the future.

Senatobia, Mississippi Junior College

This system was intended to consist of four 50 kWp arrays demonstrating different flat plate and concentrator technologies. These systems were to provide part of the Colleges's electrical and thermal energy needs. In addition, the project called for a fluidized bed biomass system to provide heat for a district type heating system for the dispersed school buildings.

The project was terminated when budgetary funds were exhausted and only part of the PV system and the biomass system had been built.

Washington, D.C., Georgetown University

This system has been designed as a 300 kWp PV system having its array integrally mounted on three sloping roofs of a specially designed intercultural building. The PV system will be interconnected with the University's electrical power system supplied by the local utility, the Potomac Electric Power Company.

Design has been completed by Hughes and component manufacturing is in process including solar cell module manufacturing. The modules will be Solarex's semicrystalline silicon cell modules. The inverter will be a DECC Helionetics unit.

Installation and initial operaton is expected to occur during 1984.

Sacramento, California, Sacramento Municipal Utility District

This project is a planned multi-phased project to provide a 100 MWp PV system for the Sacramento Municipal Utility District (SMUD). Phases one and two are intended to install a capacity of 1 MWp in each phase over the period from mid-1983 to mid-1985. Phase I has been initieated, funded and procurements executed. Modules will be provided by Arco Solar and the inverter will be provided by Windworks, Inc. Production of modules has begun. Initial specification discussions for Phase II have occurred.

Funding for the first phase has been provided by the D.O.E., California State Energy Commission and SMUD. D.O.E. funding commitments for Phase II have been obtained.

3. U.S. PRIVATE SECTOR SYSTEMS

The five systems described below are the only publicized intermediate size private sector applications. Other systems may have been built but have not been publicized.

As discussed above, the ability to fund the design and construction of this size system at current installed system costs is essentially due to benefits (subsidies) available under U.S. tax codes, Federal and state. These subsidies have stimulated considerable efforts by various PV industrial firms to carry out such projects. If these efforts are successful a number of systems may be installed over the next two and one-half years. Estimates are that such activities could result in new installations of 20-30 MWp (or more). At least 16 MW of installed capacity during this period has already been announced.

Frederick, Maryland, Solar Powered Factory

This system has been designed and installed by the Solarex Corporation. It is a 200 kWp stand-alone PV system which consists of a 200 kWp array, a 200 kW DECC DC-DC converter, two 20 kVA LeRoy Sommer motor generators and one 10 kW Abacus solid state inverter for inverting the DC to Ac, 10 DC-DC converters of various power and voltage ratings to match the requirements of production line equipment, a 2700 kWh lead-acid storage battery and a micro-computer controller capable of fully automated operations.

This system has been optimally designed to integrate with the highly energy conservative building and the low-energy, automated production line. Many special features have been included to ensure highly reliable operations with no utility back-up.

The system has been operational since October 1982 and has been undergoing a period of test and system improvement. Principle problems which have been corrected include inadequate design for module fastening on the array resulting in the loss of a few modules during high winds, electronic equipment failures during high current surges at turn on due to the instantaneous current flow from the large battery and an array circuit box diode failure. All failures have been corrected and the system is operating as designed to provide reliable power for the facility.

Denton, Maryland, Utility Demonstration

This plant is owned and operated by Copley Energy, Inc., which sells all electricity generated by the system to the local utility, Delmarva Power & Light Company. The system is sized at 15.7 kWp and uses Solarex's semicrystalline silicon solar modules in its array. The array is a specially designed, low-cost structure oriented into a fixed tilt configuration.

The plant has been operating successfully since it began selling its power to the utility on December 3, 1982.

Walt Disney World, Florida, EPCOT Center

This 70 kWp system was supplied by Solar Power Corporation for its parent Corporation Exxon's "Universe of Energy" exhibit pavilion at Disney World's EPCOT Center. The array is mounted on the pavilion's two acre roof.

The array operates at a nominal 330 VDC. Utility interconnection to the Reedy Creek Utility is provided by a Helionetics DECC power conditioning unit. A computer based control system operates the power system and acquires data for performance assessment.

The principle loads of the pavilion are six, 97 passenger electric vehicles, weighing 6.5 tons each and are driven by 5 HP electric motors and powered by lead-acid batteries. The PV system provides the energy for charging the electric car batteries.

Hesperia, California, Lugo Substation

In December 1982, the first 1 MW photovoltaic power plant began full operations. This plant, owned and operated by ARCO Solar, a subsidiary of Atlantic Richfield, was designed by the BDM Corporation. It is located in the high desert area of Southern California on a 20 acre (81,000 square meters) site next to a major substation of the Southern California Edison Company (SCE). The systems entire output is sold and delivered to the SCE grid under a power purchase agreement between ARCO and SCE.

The system consists of 108 two-axis trackers mounted on single pedestals. Mounted on each tracker are 256 PV modules covering an area of 95 square meters. Two sets of inverters - a 1,000 KVA Garrett AiResearch unit and two 500 KVA Helionetics DECC units - provide design redundancy. A computer driven control system operates the system and, for example, will cause the trackers to stow the arrays into a nearly horizontal position if winds reach a velocity of 48 km/hr., which is not unuual in that location. emergency back-up power to provide power to the trackers and control system is provided by two on-site diesel generators.

Known problems with the system to date include overturning of 6 trackers under high wind conditions and software problems with the Garrett inverter. Otherwise, SCE has stated that the system has been delivering energy near design estimates.

Carrisa Plain, San Luis Obispo County, California, Utility

ARCO Solar has announced a planned 16 MW PV system that will be designed and built to interconnect with the Pacific Gas & Electric (PG&E) Company utility system. The power purchase agreement with PG&E has been signed and an engineering contract with the Fluor Corporation has been executed. The site for the system will be provided by PG&E. It is a 640 acre (2.6 million square meters) plot on the Carrisa Plain in eastern San Luis Obispo County about 50 miles from the Pacific Ocean. The entire 16 MW system is planned to be installed during 1984 and 1985 with the first 6 MW planned to be installed by March 1984.

The system will be essentially the same design as the Lugo 1 MW plant except that the arrays mounted on the two-axis trackers will include mirrors to seek performance enhancement.

4. CONCLUSION

The U.S. government and private sector system applications described cover a wide range of system sizes, technologies, configurations, applications and environmental operating conditions. Both design and operating experience to date has been extremely beneficial in aiding designers and manufacturers alike to understand how to design and build systems which will be cost- effective for large scale PV application. Particularly useful will be the operating experience which is beginning to be gained with central power systems for utilities. This experience should enable utilities to understand the economic and operating characteristics of PV systems so that they can begin to include PV systems in their long range plans. Also, this experience will greatly aid the engineers and manufacutreres to provide the components and systems which will meet utility needs.

REFERENCES

1. Macomber, H. L., et al, "Photovoltaic Applications - Past and Future," The Conference Record of the 14th IEEE Photovoltaic Specialists Conference, January 7-10, 1980, San Diego, California, and Hesse, J. L., "Survey of Residential and Intermediate Photovoltaic Application," Proceedings of the third E. C. Photovoltaic Solar Energy Conference, October 27-31, 1980, Cannes, France.

2. Sandia Report No. SAND 83-0603, "Fourth Photovoltaic systems Definition and Applications Project Integration Meeting," April 12-14, 1983, Albuquerque, New Mexico, U.S.A.

DATA MONITORING AND PROCESSING

F.C. Treble
Consultant to the CEC
43, Pierrefondes Ave.
FARNBOROUGH, HANTS.,UK

SUMMARY

An important objective of the CEC Photovoltaic Pilot Projects Programme is to gather information which will lead to the optimisation of future pv systems in terms of performance, reliability, durability and cost.

This paper, an up-date of previous documents, sets out the monitoring requirements regarded as essential if this aim is to be achieved and makes proposals for the centralised collection and analysis of the data, leading eventually to a published report.

1. INTRODUCTION

The aims of the proposed monitoring of the CEC Photovoltaic Pilot Plants are :-

- To collect information on the performance, efficiency, reliability and lifetime of the systems and components over years of operation in the field.
- To collate and analyse these data so that the various technical options can be compared.
- To publish the results in a form suitable for the future guidance of the designers and potential users of pv systems.
- To gather on-site meteorological data which can be compared with the values used in designing the system and lead to the refinement of future designs.

These objectives were confirmed at the Contractors' Meeting in Brussels on 20-21 April, 1982 and it was accepted that, if maximum benefit was to be derived from the programme at minimum cost, the data should be centrally processed and analysed and a generally accessible report published afterwards.

In this paper, the basic monitoring requirements set out in a previous paper (1) are amended in accordance with suggestions made at the Contractors' Meeting in Crete on 14 September, 1982 and proposals are made for the subsequent analysis of the recorded data.

2. DATA RECORDING

The essential parameters to be monitored are listed in Table I, together with symbols, units and accuracies. Apart from the last item, all are hourly values, preferably derived from measurements taken at 1 minute intervals. However, if a 1 minute sampling interval has not been budgetted for, it will be acceptable to use instantaneous samples taken at 10 minute intervals.

The data should be recorded or copied off-line in a standard format on

(a) an ECMA 46 magnetic tape cassette <u>or</u>

(b) an IBM-compatible magnetic tape (9-track, 800 or 1600 bits/inch) <u>or</u>

(c) a DEC-compatible floppy disc (RX02).

3. STANDARD FORMAT

The suggested standard format is as follows :-

The results of a measurement period (1 hour if 1 minute samples are taken or 10 minutes if no averaging is used) shall form a <u>data group</u>. Each data group shall be divided into <u>records</u>, comparable to lines in printing. Records shall be separated by a carriage return (CR = ASCII(13)). Each record shall comprise individual <u>fields</u>, separated by a comma. The length of the record shall not exceed 72 characters.

All data shall be written in ASCII code. Since the comma ("," = ASCII(44)) is used as a field separator, all data can be presented in free-field format and there is no need to keep leading zeros or spaces in signed integers or decimal fractions. The full stop ("." = ASCII(46)) shall be used as the decimal point. The "E" notation, written as a mantissa followed by a characteristic (power of ten) is not permitted.

Table I <u>RECORDED PARAMETERS</u>

Item No.	Parameter	Symbol	Units	Accuracy
1	Insolation, global (horizontal)	IG	$kWhm^{-2}$	Calibrated instruments
2	Insolation, array tilt angle	IA	$kWhm^{-2}$	
3	Maximum wind speed at 10m height	VM	ms^{-1}	$\pm$ 2 ms^{-1}
4	Wind direction at 10m height	VD	deg. from N	$\pm$ 10°
5	Wind loading on modules	WLi	Pa	$\pm$ 5% f.s.
6	Mean ambient temperature in shade	TA	deg. C	$\pm$ 2°C
7	Mean module temperature (sheltered)	TC	deg. C	$\pm$ 2°C
8	Array output / MPPT input energy	EA	kWh	$\pm$ 2% f.s.
9	MPPT output energy *	EC	kWh	$\pm$ 2% f.s.
10	Battery input energy *	EBI	kWh	$\pm$ 2% f.s.
11	Battery output energy *	EBO	kWh	$\pm$ 2% f.s.
12	Inverter input energy *	EII	kWh	$\pm$ 2% f.s.
13	Inverter output energy *	EIO	kWh	$\pm$ 2% f.s.
14	Energy to (+) and from (-) grid *	ER	kWh	$\pm$ 2% f.s.
15	Energy from aux. or back-up generator *	EG	kWh	$\pm$ 2% f.s.
16	Energy to each load	ELi (i = 1 to 10)	kWh	$\pm$ 2% f.s.
17	Timing of system switching events *			$\pm$ 30 min.

* Where applicable

Each data group shall start with a <u>header</u> in the following form :-

% Station % Date,Time, Comments

The station or site of the plant may be given in full but only the first four letters are mandatory. The date specifies the day of the measurement in the form dd-mm-yy. Here, leading zeros must be written. The time of recording, in the form hh.min., is the time at the <u>end</u> of the hour in which the measurements were taken (or the instant of recording if no averaging is used). In the case of hourly data records, the space allocated for comments may be used to record the times of switching events, e.g. switching from battery to load or switching in paralleled inverters (Item 17 in Table I) in a code devised by the installation manager. Other comments may be included at his discretion.

Thus, two examples of headers are :-

% ALICUDI % 03-05-84,15.00

% FOTA % 25-03-85,13.00,EXTENDED RUN

The header shall be followed by up to 9 data records, each distinguished by an initial number. The symbols given in Table I shall be used. Data Record 1 shall contain the meteorological data (Items 1-5), Data Record 2 the temperature and global energy data (Items 6-15) and Data Record 3 the energy to each load (Item 16). Up to 10 loads can be accommodated. Data Records 4-10 are at the free disposal of the installation manager to use for additional data of his choice or leave blank, as he wishes.

Thus, the layout of Data Records 1-3 is :-

1,IG,IA,VM,VD,WL1,WL2,WL3,WL4,WL5,WL6,WL7,WL8,WL9,WL10

2,TA,TC,EA,EC,EBI,EBO,EII,EIO,ER,EG

3,EL1,EL2,EL3,EL4,EL5,EL6,EL7,EL8,EL9,EL10

If any field in the format of a record is not applicable or data are not available for it, it shall be indicated as an empty field by its final comma following directly the preceding one. However, all commas directly preceding the end-of-record, CR, shall be suppressed. Thus, if, in Data Record 3, only the 4th and 6th items are available, the record shall be written :-

3,,,,EL4,,EL6 CR

4. INSTRUMENTATION

The pyranometer used for global insolation measurements should be of a modern type (e.g. Kipp and Zonen CM10 or Eppley PSP), calibrated against a Primary Absolute Cavity Radiometer (PACRAD), whose accuracy of calibration is directly traceable to the US National Bureau of Standards. Insolation in the plane of the array may be measured by another instrument of the same type or, alternatively, with a silicon cell radiometer, such as the Haenni Solar Meter 118 or the KBE Solar Sensor SS100, similarly calibrated. The CEC (DG XII) is making arrangements for checking the calibration of all radiometers in the programme at a suitable site later this year (1983). Any unacceptable discrepancies should be eliminated by an appropriate adjustment of the calibration.

It is suggested that wind loading measurements be made by calibrated strain gauges attached to the main members of the supporting structure at points of maximum stress. Measurements should be made at three different points - front row, back row and middle of the array field.

Module temperature should preferably be measured by lightweight sensors attached to cells in specially instrumented modules. Alternatively, the temperature may be derived from the open-circuit voltage and the irradiance, following a preliminary calibration. Attaching sensors to the back of the module is the least satisfactory method.

The hourly energy parameters may be obtained by direct measurement or from voltage and current readings.

The calibration of all sensors and instruments should be checked before installation and at yearly intervals thereafter.

5. FAULT DETECTION

Means should be provided in the control room for identifying a sub-array or other component that has developed a fault. A portable test set or other means should be provided for quickly locating a faulty module.

6. LOG BOOK

To keep track of the reliability and ease of maintenance of the plant, the following events, with dates, should be recorded in a log book :-

- Results of acceptance tests and any repeat tests carried out afterwards.
- Calibration checks on instruments.
- Failures, with remedial action taken.
- Accidents, with remedial action taken.
- Severe weather (hailstorms, snowfalls).
- Changes in array orientation.
- Cleaning and other maintenance operations.

The log book should also contain a record of the outcome of investigations into failures and defects.

7. DATA ANALYSIS

It is proposed that, at suitable intervals after the completion of each plant, the contractor sends his tapes or floppy discs for the preceding period to JRC, Ispra for processing and analysis. These data should be accompanied by details of any failures, accidents, replacements, orientation changes, severe weather, etc. extracted from the log book.

JRC should compute the following graphics and parameters from the stored data :-

7.1 Variation of the mean monthly global insolation ($kWhm^{-2}$ per day) and the mean monthly insolation at the array tilt angle.

7.2 Cumulative frequency curves of the mean hourly irradiance at the array tilt angle i.e. irradiance plotted as a function of hours/day for each month of the year.

7.3 Variation of the following parameters on a daily basis :-

- Mean ambient temperature in the shade.
- Mean module temperature.
- Maximum wind speed at 10m height.
- Maximum stresses on array structure.

7.4 An energy/time histogram showing the following mean monthly parameters in units of kWh/day :-

- Load requirement
- Net output from the pv system
- Back-up inputs, where applicable
- Excess energy outputs to grid, where applicable

7.5 An energy/time histogram showing the following mean monthly parameters in units of kWh/day :-

- pv array output
- losses in the MPPT(s), battery and inverter(s), as applicable.

7.6 The following monthly and yearly energy efficiencies :-

$$\text{Array efficiency} = \frac{\text{Array output}}{\text{Insolation x array area}} \times 100\%$$

(The array area in this calculation should be the total area, excluding spaces between blocks of modules).

$$\text{MPPT efficiency} = \frac{\text{MPPT output}}{\text{MPPT input}} \times 100\%$$

$$\text{Battery efficiency} = \frac{\text{Battery output}}{\text{Battery input}} \times 100\%$$

$$\text{Inverter efficiency} = \frac{\text{Inverter output}}{\text{Inverter input}} \times 100\%$$

$$\text{Array tilt efficiency} = \frac{\text{Insolation at array tilt}}{\text{Global insolation}} \times 100\%$$

The monthly efficiencies should be presented as plots on a time base over the year.

7.7 The ratio k, defined as :-

$$k = \frac{\text{Rated power of the array (kWp)}}{\text{Average pv power output over the year (kW)}}$$

$$= \frac{\text{Rated power of the array (kWp)} \times 8760 \text{ (h)}}{\text{Total pv energy output over the year (kWh)}}$$

7.8 When the electricity is being used for a specific purpose, a plot showing the variation of the monthly output (e.g. volume of water pumped, disinfected or desalinated, volume of hydrogen produced) in the course of the year.

8. REPORTS

It is envisaged that, after the first year or two of operation, a CEC report will be prepared for publication, comparing the performance and reliability of the pilot plants on the basis of the received data and drawing conclusions on the various design options for the future guidance of the designers and potential users of pv systems. This report should also, it is suggested, address the question of cost and compare the actual and projected array and balance-of-system costs for the plants.

REFERENCE

1. F.C. Treble & G. Blaesser — Basic Requirements for Data Monitoring and Recording. Paper sent to all Pilot Project Contractors by Dr. W. Palz of the Commission of the European Communities (DG XII) on 18 May, 1982.

PROJECTS

300 kW photovoltaic plant Pellworm - Power supply for the recreation centre - Island Pellworm, Schleswig-Holstein, Federal Republic of Germany

Kythnos photovoltaic power plant - Kythnos Island, Greece

Tremiti desalination plant - Tremiti Islands, Italy

Powering of a solar heated swimming pool - Chevetogne, Belgium

Rural electrification in French Guyana - Kaw, French Overseas Department La Guyane

Power supply for TV and FM emitters - Mont Bouquet, France

Nice airport survey and control - Nice, France

The Fotavoltaic project : A 50 kW photovoltaic system to power - A dairy farm on Fota Island, Cork, Ireland

Solar-wind project Terschelling - Terschelling Island, The Netherlands

Aghia Roumeli electricity supply to an isolated village - Crete Island, Greece

Water disinfection system and cold store - Giglio Island, Italy

Solar plant for a remote Corsican village - Paomia, Rondulinu, Corse, France

Photovoltaic power plant for hydrogen production and water pumping - Hoboken near Antwerp, Belgium

Marchwood project 30kW - Marchwood Power Station site near Southampton, United Kingdom

300 kW PHOTOVOLTAIC PLANT PELLWORM

POWER SUPPLY FOR THE RECREATION CENTRE

Location	: Federal Republic of Germany Island Pellworm, Schleswig-Holstein
Power	: 300 kW peak
Contract number	: ESC P056-D
Duration of contract	: November 1981 - June 30, 1983
Total budget	: 4 655 190 ECU
CEC contribution	: 1 929 000 ECU
Other funding bodies	: - Ministry of Research and Technologies of the Federal Republic of Germany - Schleswag AG
Project leader	: Hans-Jürgen Lowalt
Contractor	: AEG-TELEFUNKEN A471V21 Industriestrasse 29 D - 2000 WEDEL Tel. : 04103-702 786 Telex : 2 189 520
Authors	: Hartmut Green Hans-Jürgen Lowalt Bernd Proetel Peter Thoel

SUMMARY - INTRODUCTION

The photovoltaic project Pellworm with a peak power of 300 kW is described. It is located in the northern part of the Federal Republic of Germany on the island of Pellworm in the North Sea.

The converted electrical energy is to supply the recreation centre of Pellworm in an independent mode. Surplus energy is to be fed into the grid. The seasonal power demand of the consumers in the recreation centre are well matched to the amount of converted electricity, this means high energy needs only in high season (summer).

For testing reasons one half of the module structures are made of tropical hardwood material, while the other half is of hot galvanized steel. The structures are built up in a way that the site can be used for grazing sheep. Overall 17 568 modules are mounted. 20 multicrystalline silicon cells connected in series and glass encapsulated makes up one module. All modules are fixed with a tilt angle of 40°.

To guarantee the independent operation mode, a battery with a four days capacity of more than 8 000 Ah is installed. To realize an operation as buffer battery and a simultaneous charging, the battery is divided into two operational parts. The grid connection is possible for the consumer as back-up energy supply.

Energy transmission to the recreation centre is done by step up transformers to reduce cable losses on the 1 300 m cable run.

As a superimposed control- and monitoring system a micro computer is integrated.

2. GENERAL TECHNICAL DESCRIPTION

2.1 Data

Rated Peak Power	300 kW
Nominal DC-Voltage	346 V
Solar Generator Field	
Overall Area	28 000 m^2
Structures Area	16 500 m^2
Module Area	4 500 m^2
Building	15 m x 10 m
Battery capacity (10 h)	6 000 Ah (long term discharge see fig. 1)
Self Commuted Inverters	2 x 75 kVA
Output Voltage	3 x 380 V $\pm$ 2%
Frequency	50 Hz
Total Harmonic Distortion	3% max. under full load
Line Commuted Inverter	450 kVA
Output Voltage	3 x 380 V
DC = Input Current	1 000 A
Transmission Line	
Length	1 300 m
Voltage	20 kV

2.2 Solar Generator

The generator field with the peak power of 300 kW is installed on 61 structure rows, of which 29 rows are made of tropical hard wood and 32 rows made of hot galvanized steel structures. To control the 300 kWp, the solar generator is divided in 21 switchable groups. 48 modules connected in series performing one string with the nominal voltage of 346 V DC. 6 strings connected in parallel on the structures and fed to the DC-switchboard in the building by cables of different cross sections, depending on their length.

2.3 Battery

For the energy storage lead acid batteries are used. Each individual cell has a rated voltage of 2 V, therefore 173 cells had to be connected in series to provide the necessary system voltage of 346 V. All cell connectors and poles are insulated, so that no direct contact can be made with current conducting metal parts. In the fitted recombinator plugs the gases developed through charging are reconverted to water and returned to the electrolyte. The battery is divided into two parts so that it is possible to work with one half of the battery as buffer battery, while the other half is charged in a defined way.

2.4 Inverters

For the independent operation mode of the recreation centre two self commutated inverters of each 75 kVA are used. According to the momentary load demand each of the inverters can be switched on or off, controlled by the superimposed micro processor system.

A line commutated inverter is used to feed the surplus energy to the grid or if an emergency charging of the batteries is necessary.

2.5 Superimposed Control System

The automatic operation mode is accomplished by a micro processor system. To use the converted energy in an optimal way the micro processor has to take over a lot of decisions, among others e.g. how many solar generator groups are to be switched on to the chargeable battery and to the buffer battery, what part of the battery will be used as buffer battery, is it necessary to use one or two self commutated inverters or can surplus energy be fed into the grid. All this is only possible by a very complex monitoring system so all essential values and signals of the whole plant are collected by the micro processor.

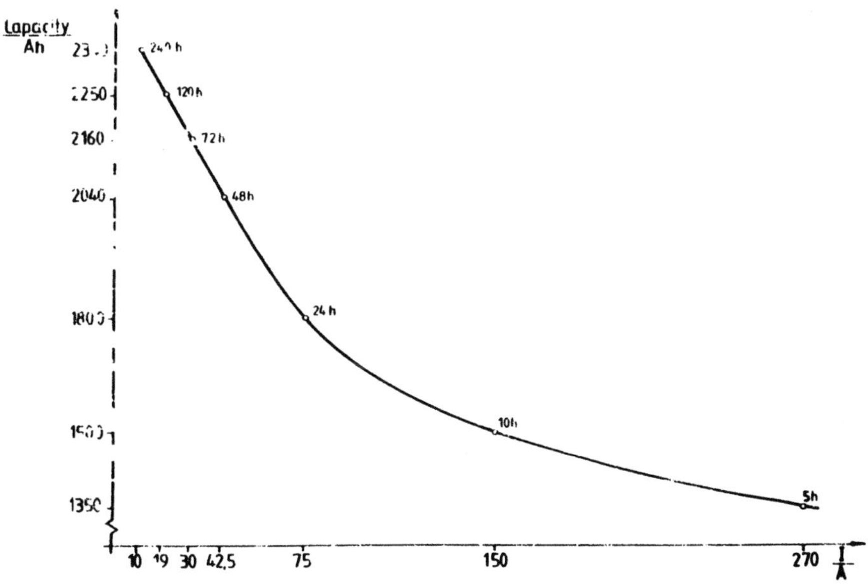

Fig. 1 - Capacity curve for long term discharge.
Nominal 10 h capacity for one battery 1 500 Ah
4 batteries are in parallel, so the nominal capacity is 6 000 Ah (10 h)

3. ARCHITECTURAL AND ENVIRONMENTAL ASPECTS

The island Pellworm located in the North Sea is flat, with its surface below sea level. The main income for the inhabitants is tourism and agriculture. So the important aspects of installing the solar generator plant was to make no alteration on the character of the environment and the possibility to use the site for grazing purposes. The building for the energy storage and the power conditining is built up in an appropriate appearance to other buildings on the island. Sheep need about 80 cm free room in height to graze. An optimum for building up the fixed structures with an angle of 40°, to enable an easy connection of the individual module, is a maximum height of about 2.5 m, by the lower row of modules in approx. 1 m height. To get a well defined layout of the plant, the structures made of different materials, are divided from each other by an approximately 30 m grass area. From the main view point, the building, an unimpeded view on the landscape is possible. In addition bushes will be planted around the whole array area.

Fig. 2 - Sheep grazing on the solar generator field

4. OVERVIEW OF NEW TECHNOLOGICAL DEVELOPMENTS

Self commutated Three-Phase Static Inverters

In this Pellworm Solar Power Plant a new generation of three-phase static inverters were implemented. Some typical characteristics of these units are :

- high efficiency also at low load operation
- short recovery time
- parallel operation of up to 4 units possible with equal power distribution
- small distribution factor
- low radio interference
- small volume

Description ref. to chapter 9.

Battery

For the energy storage fully encapsulated lead acid batteries equipped with recombinator plugs are used. To built up such batteries the standard requirements are changed. There is no longer the necessity to use explosion-proofed apparatus, no continuous operation of a forced air exchange, no water supply in the installation room and the gaps between the battery rows are minimized.
All these points are of great interest for specific costs in future solar energy application. To reach a long lifespan of the batteries they are divided into two operational banks, so one half of the batteries can be charged in a defined way, while the other half serves as buffer battery. Ref. to chapter 9.

Two Busbar System

To achieve the special operation mode of the two battery banks mentioned above, there are two DC busbars. One half of the battery is connected to each busbar. The 21 solar generator groups are operational switchable to each busbar. The DC-power supply for the self commutated inverters is switchable without interruption from one to the other bus. Ref. to chapter 9.

Micro Processor

The superimposed control system is built up by micro processors which are controlled and monitored by separate, so called, watch-dog-circuits. This ensures especially for the operation without personnel, a reliable and failsafe running. Ref. to chapter 10.

5. SOLAR GENERATOR MODULES

The solar generator is built up of AEG standard glass encapsulated modules type PQ 10/20/0 with multicrystalline silicon cells. 20 cells connected in series with a dimension of each 10 cm by 10 cm making up one module. The 20 cells are embedded in glass/PVB/glass and fitted in a stainless steel frame so that they are well suited for the installation in the very severe environmental conditions on the island.

The electrical and mechanical data may be taken from the following data sheet.

Electrical Data

The modules PQ 10/20/01 and PQ 10/20/02 are designed to charge 6 V-batteries.
Any series and parallel connections are possible.

Characteristic values (AM 1.5–100 mW/cm²)	Solar cell temperature 0 °C	25 °C	60 °C
Open circuit voltage (V)	12,3	11,2	9,6
Short circuit current (A)	2,37	2,41	2,46
Current at maximum Power (A)	2,18	2,20	2,23
Nominal Power (W)	21,3	19,2	16,2

Max. difference to nominal power less than 10%

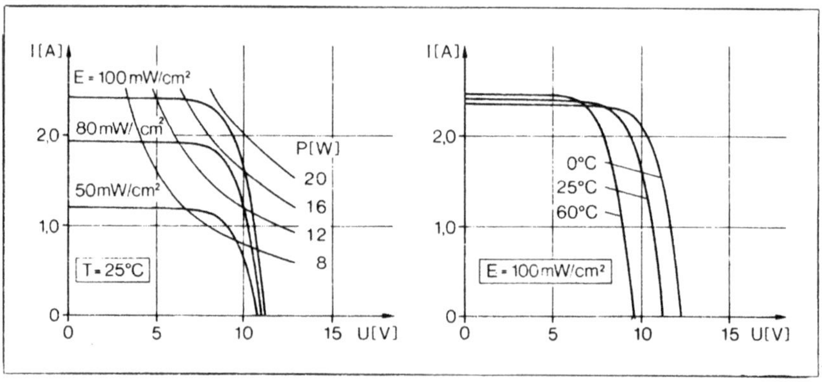

Structural Data

PQ 10/20/01
PQ 10/20/02

Solar cells
Base material:
Multicrystalline silicon

Size
10 cm x 10 cm

Encapsulation
Glass/PVB/Glass
Stainless steel frame

Weight
3850 g

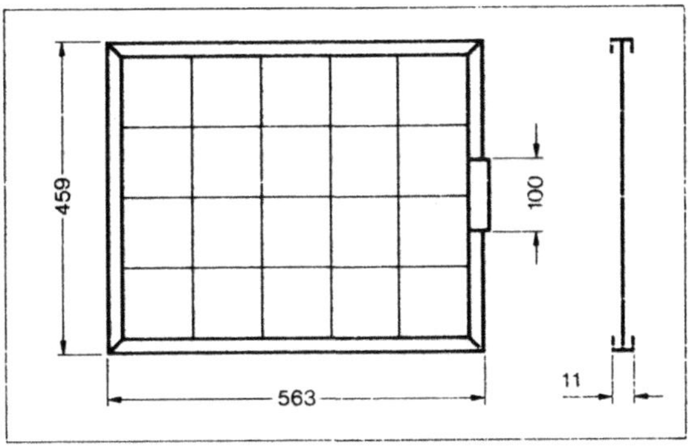

Dimensions (mm) PQ 10/20/01, PQ 10/20/02

6. ARRAY AND CABLING

The solar array is divided into 21 groups. 20 of these groups (no. 1 - 20) consist of 3 sub-groups of 6 strings each. Group no. 21 has only 1 sub-group of 6 strings.

Each string has 48 modules of AEG type PQ 10/20/0. Thus the total number of modules is :

20 groups	x 3 sub-groups	x 6 strings	=	360 strings
1 group	x 1 sub-group	x 6 strings	=	6 strings
				366 strings

366 strings x 48 modules = 17 568 modules.

The modules are interconnected electrically by cables with their single-pole connector-plugs. The beginning and end of each string is connected to the string junction box by special connector cables.

The module connector-plugs are provided to facilitate the system integration. An additional securing sleeve prevents this type of plug being opened by unauthorised persons. This securing sleeve is clamped to the module frame and thus the connecting cables are fixed additionally.

The six double wire string cables of each sub-group are conducted to a sub-group junction box (see Fig. 3). Here the 6 negative conductors are connected to the negative junction box bus directly, the 6 positive conductors by two diodes and the string fuse to the positive bus. Both busses and the protection earth are conducted to the switchboard in the battery and switch board house by a cable type NYY 3 x 16, NYY 3 x 25 or NYY 3 x 35 depending on their length.

Thus 61 sub-group cables go into the main switch board. 3 sub-group cables are connected at the switch board terminal to form one group. The sub-group cables are embedded in trenches in the frostfree zone at appr. 80 cm depth.

Fig. 3 - Solar generator sub-group connection box

6 times 48 modules in series are connected here in parallel. A disconnection of each string is possible by opening the fuses.

7. ARRAY AND SUPPORT STRUCTURES

The photovoltaic plant is installed on an area of 28 000 m^2. Loadable stratum is found in a depth of 20 m. The surrounding land is used as farm-land, mainly pasture ground, crossed by draining channels.

The climatic conditions are very rough, stormy weather with salt foggy air, may occur in all seasons. Humidity of about 80-90%. At noon the highest sun-angle is 59° in summer and the lowest 12° in winter.

All these points contributed to the design of the support structures of the PV-panels. The total height was limited due to the wind-load and also by the fact that the array must fit well into the natural background. On the other hand the ground should be continually used for agriculture and sheep grazing. The height of the lowest panel is 1 m over ground while the highest point is 2.60 m. All panels are mounted on fixed angle structures. The angle of 40° is optimized, so that a maximum average output over the whole year is obtained.

16 strings of 48 modules each are connected to one group. These 16 strings are mounted on 8 rigs. Each rig contains 4 rows of modules, each row consisting of 24 in series. Elecrically the rigs are divided in two parts. The 24 modules of the two lower rows and those of the 2 upper rows are connected in series.

To keep costs low was an essential goal for the design of the structure, especially in view of the unfavourable environmental conditions like corrosive atmosphere, high wind load and bad soil carrying capacity. At first tests were carried out to find possibility of using a minimum of fixing points and of mounting brackets for fixing the modules.

The result of the test showed that only two fixing points in the middle of the long sides of the module were necessary. Only one girder for each row of modules is needed. The construction design could be defined as an optimization of different construction materials and methods with view to a low-cost solution. The materials examined were painted steel, hot galvanized steel, stainless steel, aluminium, pine wood pressure impregnated and bongossi (a tropical hard wood). Not only from the cost viewpoint, the hot galvanized steel and bongossi wood was chosen. Hot galvanized steel is the most widely used metal in coastal regions. The lifetime of the hot galvanizing will be more than 15 years without additional protection precautions. Manufacture of steel is conventional and its static values can be calculated. The same can be stated for bongossi wood. It has an extremely long lifespan, much more than 20 years without any treatment. The costs for both materials are nearly the same.

The supporting structure of the four rows of girders were designed as triangles with a tilt angle of 40° to the south. The advantage of the triangle is that its static is defined. The calculations were based on the German standard DIN 1055, however with a wind load of 120 km/h, i.e. 0,8 kN/m^2 (rectangular area with a c_w-value 1.1).

Out of former steel constructions it is known that not only the static load has to be considered but also vibration problems resulting from varying wind speeds. After the first calculation, a test rig was built to evaluate the natural frequency. The evaluation showed that a strut was necessary to stiffen the diagonal side of the triangle. The girders were connected by a U-bar every one third between the trestles, and the diagonal wind braces were strengthened. The natural frequency of the rig was higher than the exciting frequency of the wind (in maximum 10 Hz). For the higher impulsive moments there are no problems of vibration with the wood rigs. All mounting brackets (for the modules, for the steel rig as well as for wood rig) are made out of stainless steel.

As stated, the soil conditions are bad on site. The site itself lies below sea level. The groundwater level is at a depth of about 2 m. The supporting capacity is only 8 kN/m^2 in the frostfree zone. The foundation must be a very light construction. Prefabricated $\perp$ - bars made out of reinforced concrete were utilized.

Fig. 4 - View of steel structures from behind

Fig. 5 - View of wooden structures from behind

8. LIGHTNING PROTECTION

During design phase special attention was paid to the danger of lightning strikes. Originally three steps of protection were considered :

1. Construction of a number of lightning conductors distributed over the array area.

2. Coarse electric circuitry protection by metal-oxide varistors, air gap protectors and lightning chokes.

3. Fine protection for measuring and signal circuits by surge suppressors, diodes and isolation transformers.

After more detailed evaluation, also regarding the EC paper of L. Thione, R. Buccianti and L. Dellera, "Protection against lightning of photovoltaic generation systems", Brussels, 1981, it was realized that lightning conductors in the array field need not to be installed as the probability of strikes into the field is calculated to be very low.

Therefore protection is ensured in two cases :
- lightning strikes into the air termination system of the battery house
- induced transients in the array field cables by strike in the vicinity.

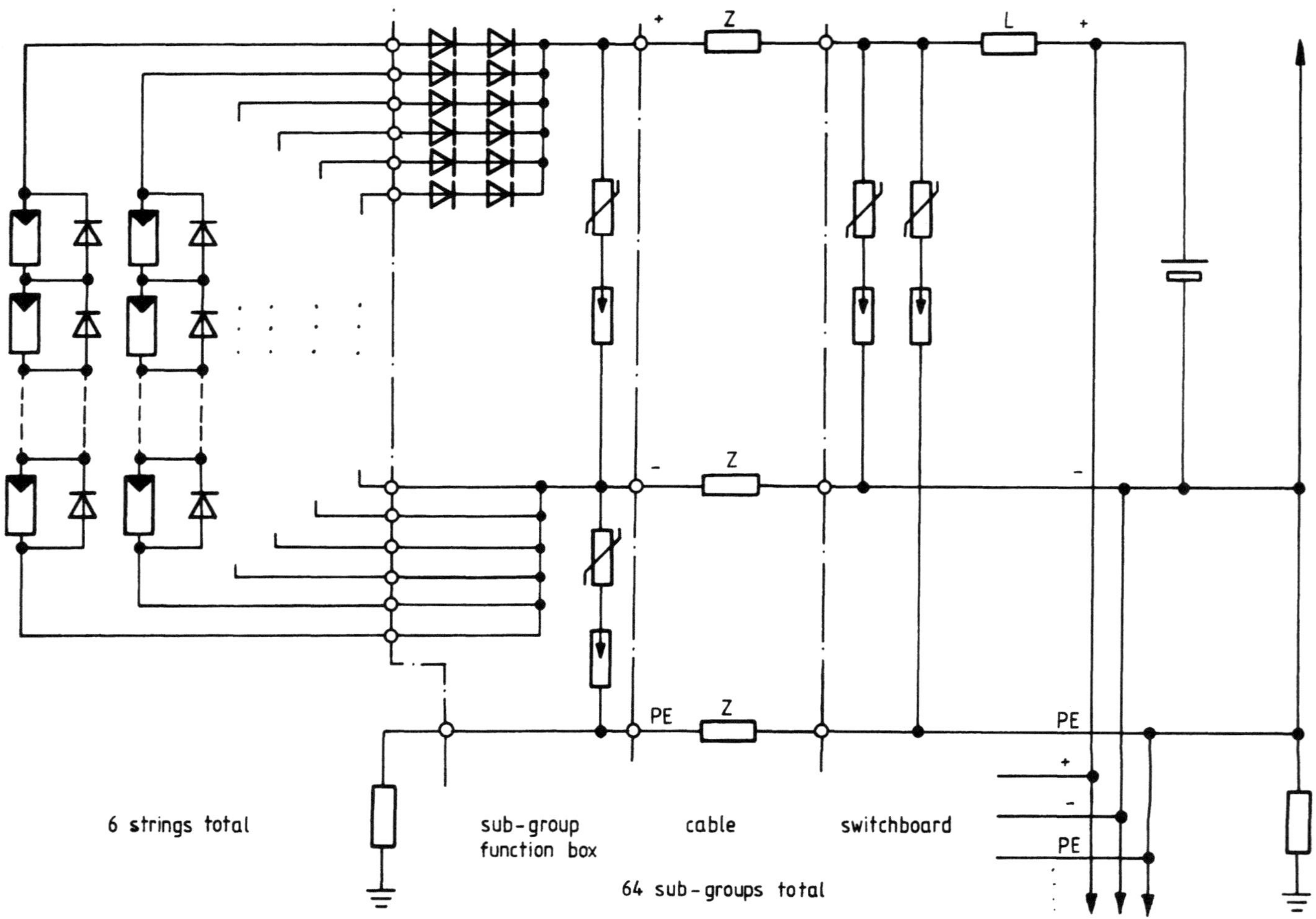

Fig. 6 - Block diagram for lightning protection

The protection system for the array sub-group cables is shown in Fig. 6. In the sub-group junction box one metal-oxide varistor with an air gap protector is provided both between the negative conductor and protection earth. In the switchboard these components are mounted between each positive and each negative conductors of the individual sub-group cables and between the positive conductor and the protection earth. A lightning protection coil in each sub-group line prevents the transients to run into the DC bus.

For the string series diode two components are provided to withstand positive voltage surges up to 2 000 V. With negative strikes into the battery house termination a varistor is shunted by the module bypass diodes.

In the switchboard a lot of components are provided for fine protection of the measurement- and control-equipments.

9. POWER CONDITIONING AND STORAGE CONCEPT

- Explanation of Plant Equipments

The general idea of the Pellworm Power Plant is to get the maximum benefit from the 300 kW solar generator.

First of all the recreating centre of Pellworm, i.e. a locally delimited consumer network has to be supplied. It shall be demonstrated that an independent operation is possible. This should be the normal operation.

On the other hand it has been found that there is surplus energy especially during the periods when the recreation centre is closed. This energy will be passed directly into the utility grid - not over the centre.

In the first case the storage batteries and the self-commutated inverters, in the second the line-commutated inverters are provided.

The maximum power demand of the recreation centre is 100 kW at a power factor of 0,8, the daily energy consumption 1 300 kWh maximum. Two inverters of 75 kVA each are installed so that the maximum power demand can be met. At low power periods only one inverter is in operation with the power loss of only one unit. The daily energy gain of the array is calculated to amount 1 700 kWh per day in June and more than 1 400 kWh per day from April to August. The data of energy generation and consumption in July 1983 have not yet been prepared but the recreation centre could be supplied from the day of commissioning up to now.

By the availability of the storage batteries feeding into the utility grid is independent from the actual insolation. The energy can be fed whenever it is needed by the utility grid. As the price of electric energy depends on the time when it is available there are promising aspects for a delivery contract with the utility board.

- Voltage level considerations

The choice of the DC voltage rating and of the number of modules per string and the number of battery cells in series depends on two values:

- The guaranteed voltage of the contactors applied is limited to 620V.
- The battery maximum charge voltage, i.e. gassing limit voltage, shall be in the near of the Maximum Power Point voltage at the junction temperature of an average sunny day.

The minimum temperature encountered is -10° C and the open circuit voltage then at full insolation is 12.8 V per module. Thus 48 modules are necessary (614,4 V open circuit).

Considering the influence of the sea at the island Pellworm - temperature and wind - an average junction temperature of 35° C during the sunshine hours in summer is encountered. The MPP voltage per module is 398 V. Regarding optimum efficiency a charge voltage level of 2.3 V per cell (0.05 V below gassing limit) was encountered for battery design, resulting in 173 cells. Thus the nominal DC voltage is 346 V. The voltage range is 320 V minimum (1.85 V/cell) up to 407 V maximum (2.35 V/cell).

- Inverters

Technical data :

Rated power	: 75 kVA each at cos phi = 0.8 ind
Efficiency at rated power	: 91,5 % (curve see fig. 7)
Rated input voltage	: 346 V DC
Rated output voltage	: 380 V, 3 ph
Rated frequency	: 50 Hz
Recovery time at 100% load changes	: 20 ms max.

The typical characteristics as shown in chapter 4 are achieved by a unique combination of inverter power circuitry and control technique : the step pulse technique.

These inverters have two three phase inverter stacks which are connected in parallel with respect to the DC side (see fig. 8). Related to the AC side these stacks are connected in series but with an electrical phase shift of 30°. The commutation circuit applied is of the McMurray type. The main thyristors are controlled in such a way that each inverter branch produces two power pulses in each period. They are combined in the main transformer. The output voltage is sinuoidal composed of squares of 30° el. duration: Thus the content of harmonics distortion is low and only a minimum of output filtering is necessary : a simple low-pass filter. The series induction of this filter is integrated into the main transformers.

The control circuitry is provided for parallel operation of two inverters. It allows each inverter to be started and stopped at any time. Actually these start and stop signals are generated by the monitoring system according to an efficiency optimisation program.

Additionally each inverter has its own monitoring unit. A number of sensors are distributed over the circuitry. Thus each operation status can be displayed. If any failure is detected the inverter is switched off automatically before any damage occurs. The monitoring unit stores all incoming failure signals accordingly in chronological order. These signals can be read out by service personnel. The failure signals and its order gives useful hints for correction.

As inverter for the grid feed in, a line commutated type with two six pulse thyristor bridges in parallel is used. This inverter is to operate in max. 1 000 Amps DC at an output voltage of 380 V 3 phase.

In the reverse direction an emergency charging for eventually battery reconditioning is possible.

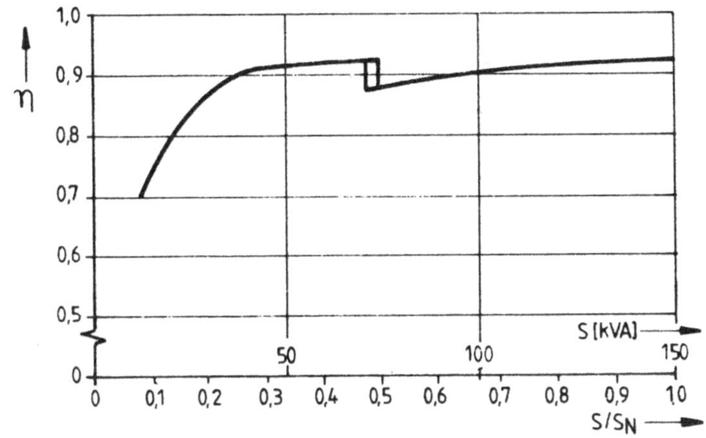

Fig. 7 - Efficiency curve of the self commutated inverters.
According to the momentary load demand two inverters are switched in parallel

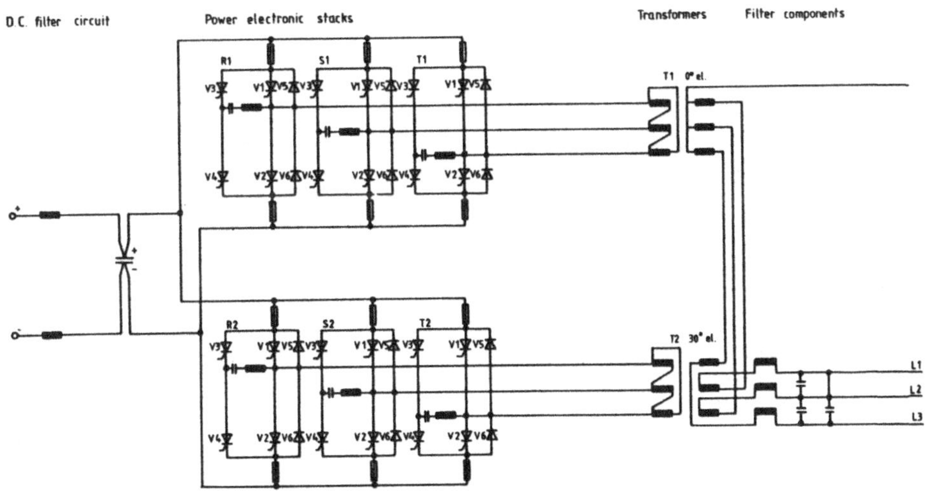

Fig. 8 - Self commutated inverter
Basic circuit diagram

- Storage Concept

During the design phase the question was discussed whether the solar array should feed the battery directly or by insertion of a maximum power point tracker (MPPT) i.e. a DC/DC converter which shifts the operating point of the solar array always into the maximum power position. Computer simulations showed that the additional energy gained by MPPT is about 5 % over a well matched direct feeding system. From this figure the power loss of the MPPT has to be deducted. On cloudy days there is no energy gain by the MPPT.

On account of this, AEG decided to use a direct feeding system. The total array is divided into array groups of appr. 15 kWp. If the battery is discharged all available groups feed into the battery. The battery voltage rises with increasing charge rate. However, the gassing voltage limit shall not be exceeded to avoid water loss in the battery cells and damage of the equipment connected by overvoltage. Therefore the battery voltage is kept constant once the limit is reached by a successive switching over of the array groups. When the insolation decreases or the load increases the groups are switched back again, (ref. to Fig. 9 : Functional block diagram of the power plant).

By this technique a minimum of power losses are generated in the switchboard.

- Two Busbar Systems, two Batteries

The long time lead battery test on battery performance in charging procedures typically for photovoltaic system application have shown that rapid degradation occurs if batteries are not charged completely at periodic intervals. At photovoltaic plants with only one battery operating permanently in floating mode complete charging is not posible even if enough energy is available. Full charging needs 12 hours minimum at a continuous charge rate. This is not possible if some current has to be fed to the consumers. During the night the battery is discharged and during the next and consecutive days the battery is not fully charged again.

Therefore in Pellworm two batteries are provided. While one of them is operating in the floating mode, the other is available for reconditioning. Thus an optimum liftetime is ensured. Each battery has its own DC bus - the double bus system. Each one of the solar array, i.e. generator groups and each of the DC consumers can be connected with one of the two busbars alternatively.

Each one of the 22 array groups (including one spare group) has 2 contactors coordinated. Each group can be connected to bus no. 1 or bus no. 2 or be switched open. In this way the charge regulation by switching array groups is possible for both batteries.

Both DC busbars are split into two parts by the battery breakers of double pole type. The generator part of the bus and the consuner part are connected to the battery by one breaker pole each. If the breaker has cut off the two parts are disconnected from each other. Thus the array opencurcuit voltage cannot be transferred to the inverter input terminals.

The line commutated inverter, by the way, is connected to the generator side of the busses. This inverter can also operate without battery connection. (ref. to Fig. 9).

Both the inverters for the recreation centre and the auxiliary inverter have their own switch-over arrangement consisting of two power contactors with diodes in line. These contactors are controlled overlapping. One contactor connects the bus to be switched on before the bus to be switched off is disconnected. This procedure ensures continuous inverter operation during switch-over procedure. Compensating currents from one bus to the other are blocked by the series diodes. These diodes are bridged after the change-over procedure at the feeder for the self commutated inverters. This is to eliminate the power losses of these high current diodes.

10. CONTROL- AND MONITORING SYSTEM

10.1 Computer Hardware

The block diagram of the computer system is shown in Fig. 10. Two 8 bit micro-cumputer systems based on INTEL 8085 are used which serve different functions. Communication between these computers is possible via a serial asynchronous RS232-link.

Communication computer (μCI)

The following peripheral equipment is connected to this unit :

- CRT terminal (main terminal)
  serves as the interface to the operating personnel

- Magnetic Type Cartridge Drive
  used for data recording

- Matrix impact printer
  used for permanent read out of data and messages.

Control computer (μCII)

This computer is used primarily for controlling and monitoring all power electronics. In order to do this autonomously all data acquisition is done by this unit, too. Therefore, extensive analog and digital input/output parts are connected to the processor.

The interprocessor link is used to transmit process parameters and commands from μCI to μCII and in turn to transfer acquired data for further processing back to μCI.

Safety Concept

Since the monitoring system must operate without supervision by operating personnel, special precautions had to be taken to insure failsafe operation. Especially the battery system is very sensitive to incorrect

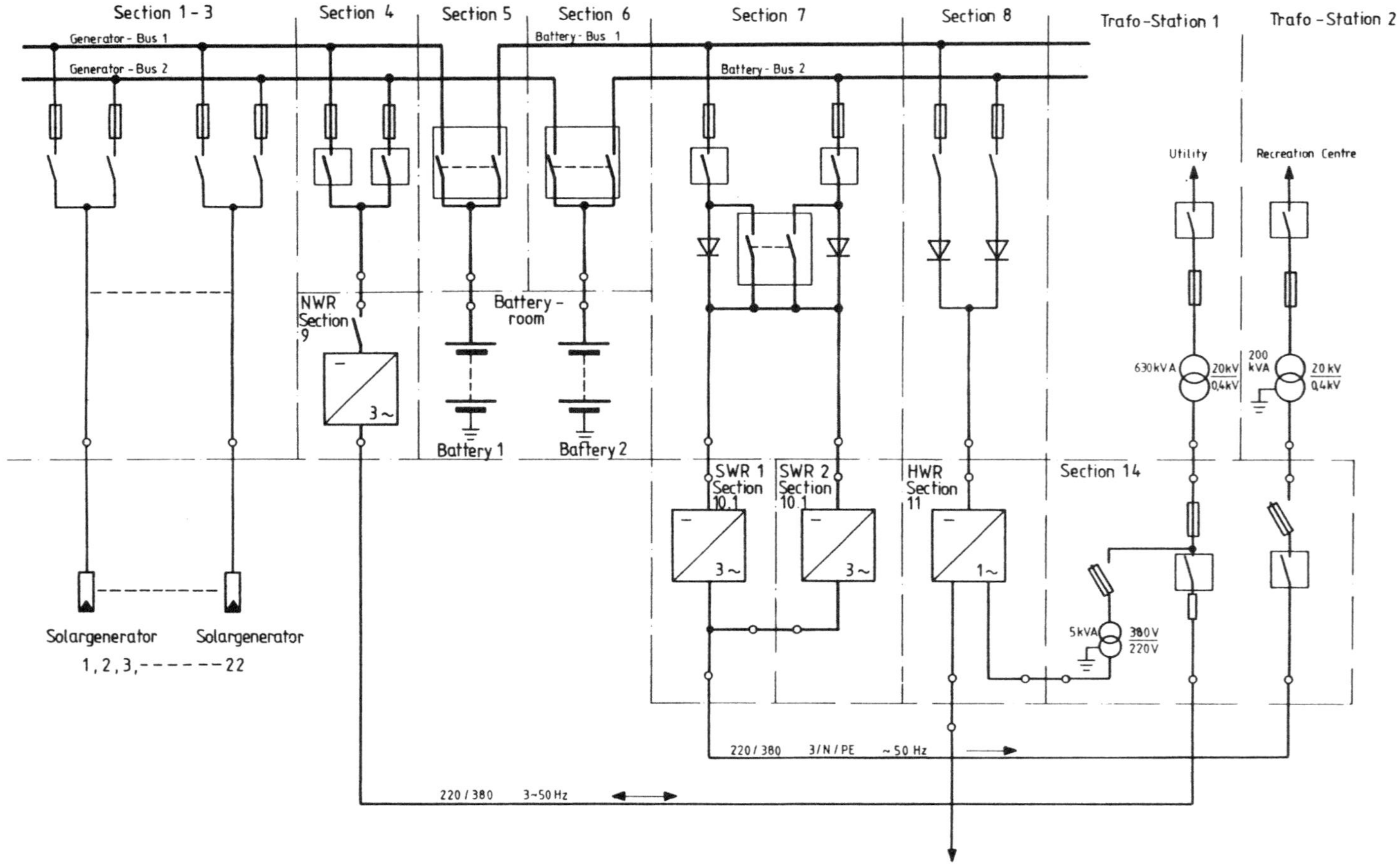

Fig. 9 - Functional block diagram of the power plant

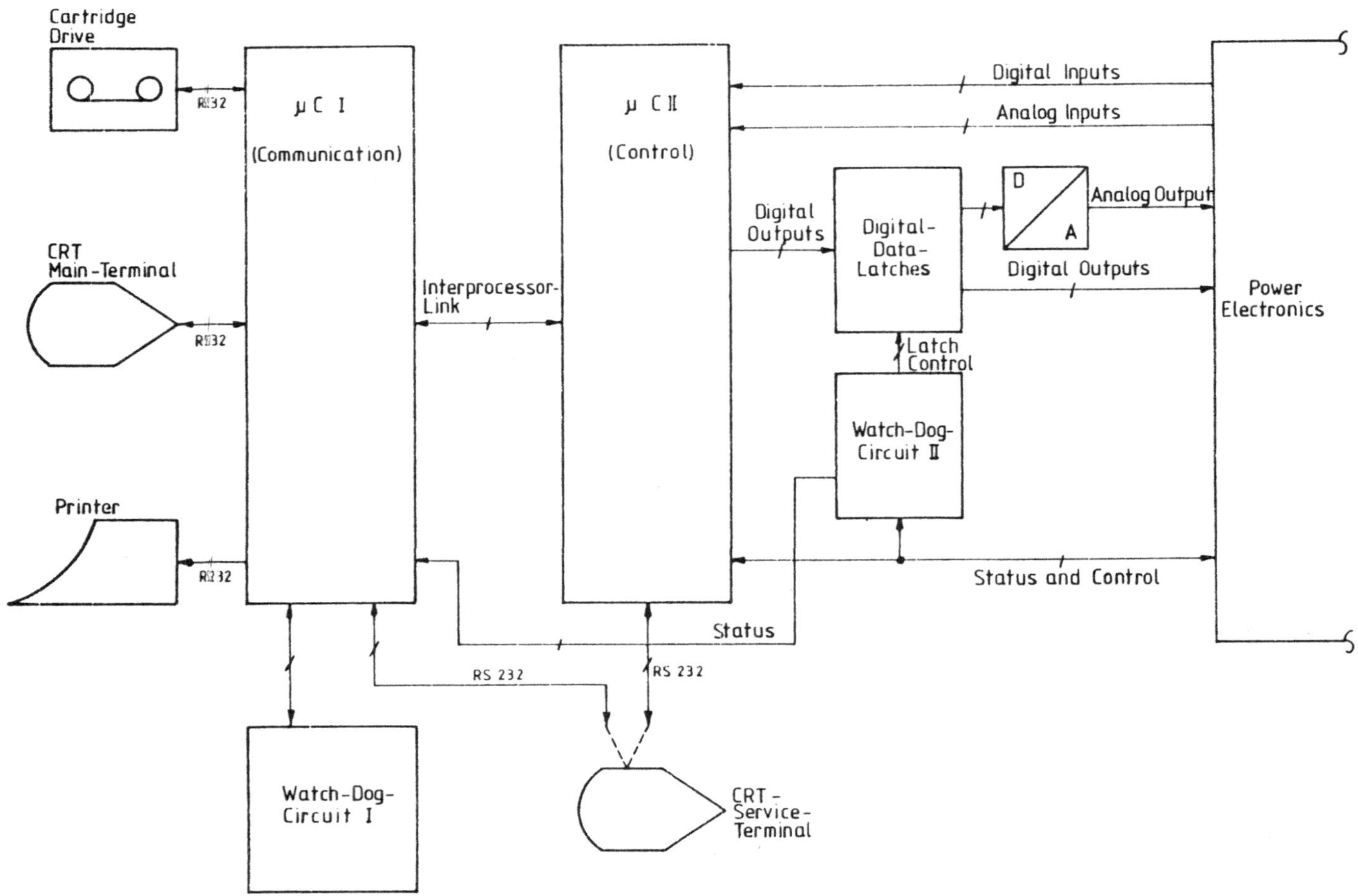

Fig. 10 - Computer system Block diagram

charge/discharge regulation. Therefore, a combination of hardware and software measures are implemented in the system to reduce the risk of fatal errors. As far as hardware is concerned each computer is controlled by a separate watch-dog-circuit. Whenever computer μCI does not transmit periodically a control signal to watch-dog I it is assumed that a malfunctioning has occured. Subsequently a cold start, i.e. starting from the beginning of the program is initiated.

Random interferences like EMC do not lead to a permanent break-down.

For the control computer μCII a similar watch-dog-circuit is used, but instead of performing a cold start, the last state of the system is stored in specially built digital data latches. As soon as the interference has disappeared a so-called warm start is performed, i.e. continuing from the last state. Only in case that several successive warm starts are performed the complete system is turned off.

A further step to reduce risk of malfunctioning is by redundancy of certain sensors which are installed twice. These output signals are compared by software and are also checked for plausibility and limits.

10.2 Computer Software

Overall Software Concept

As explained above a double microcomputer system is used for the monitoring and recording function. Both computers are running under the control of a real time multitasking operating system. These operating systems coordinate the tasks which are running on each computer and also manage the interprocessor communication via a so-called mailbox function.

The overall software structure is shown in Fig. 11. The main tasks are :

- Control and power management

- Supervision of the power electronics and the batteries

- Communication consisting of the interprocessor communication and the interface to the human operator

- Monitoring, i.e. displaying of relevant power plant data and status information and recording on magnetic type cartrige.

The photovoltaic power plant can be operated in either automatic or manual mode. In automatic mode control and supervision, i.e. the essential functions for safe operation are performed completely by the control computer.

For all other tasks both computers are needed and operate in an interactively.

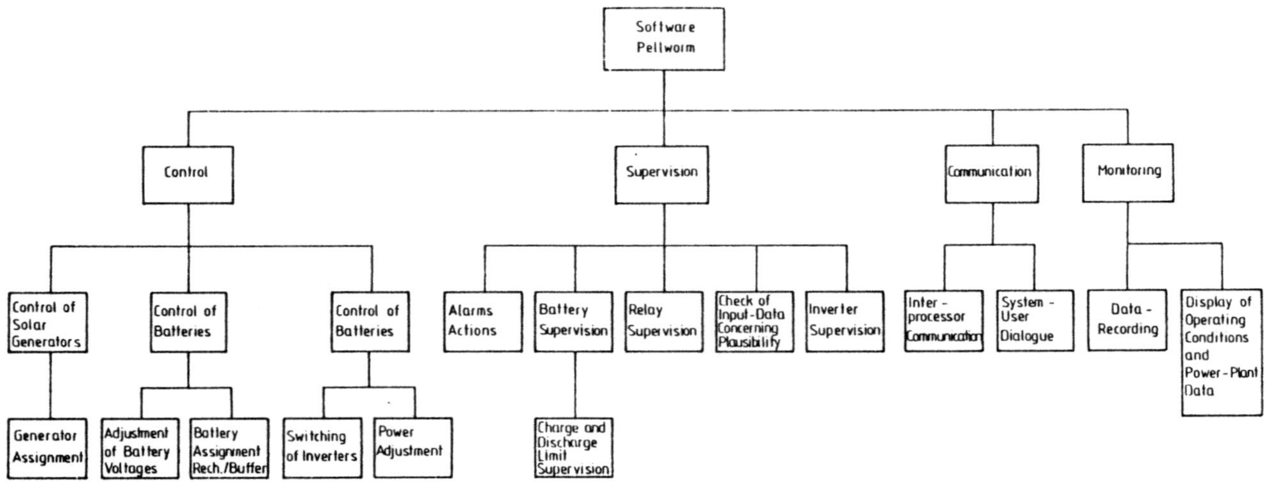

Fig. 11 - Computer software Block diagram

Control processor software

The main control functions are also shown in Fig. 11. Since not all details can be covered in this paper only the key-items are briefly listed below.

a) Solar generator control

- Monitoring the switched states of the 21 generator strings

- Measurement of generator current, voltage and temperature

- Counting the number of switch cycles for each string

- Averaging of measured values

- Switching of strings under program control or upon manual command entry via keyboard. The strings can individually be connected to the buffer battery, the recharge battery or can be turned off or blocked.

b) Battery control

- Assignment of the buffer and recharge battery to the two DC busses either upon manual command or under program control depending on a combination of different criteria

- Checking of battery voltage taking into account impacts caused by battery temperatures, charge- and discharge current

- Control of charge current

c) Control of inverters

The two self-commutated inverters and the line commutated inverter can be turned on and off either under program control or upon manual command depending upon the load current from the recreation centre and available power.

d) Control of inverters

It is possible to switch back and forth between these two modes.

In manual mode monitoring of system status and data acquisition is continued without interruption. This allows or switching back to automatic mode at any time.

Communication processor software

A real-time-multitasking operating system coordinates several tasks which are run on the communication processor.

The main task is a dialog-oriented program by which the user can select in a so-called menu-type operation several functions :

- Display of system status

- Switching back to manual control

- Display and modification of system component status (solar strings and inverters)

- Display and switching of batteries, i.e. assignment to bus 1 or 2

- Display of system data

- Remote control of data recording

- Setting of the real-time-clock.

For the data display function a submenu is used by which a selection can be made between different system data :

- Battery data

- Solar generator string data connected to the buffer battery

- Solar generator string data connected to the recharge battery

- Power data, i.e. input and output power of the different inverters.

Most data momentary as well as averaged values are displayed.

All those functions which only comprise displaying of data or status informaton can be accessed by every user. Functions which result in control or switching of system components are only accessable to authorized personnel. To achieve this, a number of passwords are available. After starting the dialog program the individual password has to be entered at the keyboard and if entry is correct the password number (not the password itself) and entry time are printed out on the printer.

The printer is used for all messages which should not be lost, i.e. mainly error messages which are important for service and maintenance. Error messages contain information about type, location (system component as far as possible) and time of occurrence.

Data recording is another task which is controlled by the communication processor. About 60 different data are recorded periodically (10 minutes interval) as one block on tape. Among these data are

- Meteorological data

- DC bus currents (averaged values)

- Number of strings connected to each bus

- Average output power

- Battery voltages and temperature

Recording time on one cartridge is 32 days.
Evaluation of data will be carried out off-line on a different computer system.

11. SYSTEM INTEGRATION AND INSTALLATION

The installation of the whole plant took only from August 82 until the end of June 83. All technical hardware were mounted and installed on well planned and prefabricted devices. It is worth mentioning the concrete foundations in field, the steel- and wooden structures, all the cable work and the house building. The switch cabinets AC and DC, inverters, micro-processor, batteries and modules were planned to be ready for delivery a few months before the final contract date June, 30th 1983. So for integration and function test enough time remained. The mounting-, wiring- and connection-work of the modules is simple and fast, so that additional special efforts on the development of modules is not necessary. During the mounting and transportation of the installed 17 568 modules, only 16 modules (0,09%) had to be replaced because of broken glasses or wiring- and plug failures.

CONCLUSION AND FIRST RESULTS

Since the beginning of July 83 the plant has supplied the electrical energy for the recreation centre in a stand-alone operation.mode. Although the plant is running the very first operation phase, no problems by supplying energy, due to the redundancy of all important plant equipment was experienced. Already this short operation time is showing that a few changes might be necessary for the very optimal use of all converted sunlight energy. But with the use of a micro processor for all control functions, this can be done easily by software modifications.

It can be stated, that the whole equipment is already running most satisfactorily and after a well observed period of operation the important knowledge required to build up and operate photovoltaic plants in an optimal way also up to the Mega Watt power range will be achieved.

KYTHNOS PHOTOVOLTAIC POWER PLANT

Location, country:	Kythnos Island, Greece
Rated peak power of array:	100 kWp
Contract number:	ESC-P070D(B)
Contractor:	Consortium Siemens - PPC - Varta
Duration of contract:	until 30 June 1983

ABSTRACT

The Kythnos Photovoltaic Power Plant has been installed on the Greek island Kythnos which is located about 60 km south of Athens. The aim of this project is supplying electricity together with the existing diesel power station to the island consumers and saving of high costly fuel oil. A photovoltaic solar generator of 100 kW peak converts solar energy directly in electrical energy. To feed into the existing island grid with a 3 phase AC system a special system technology was developed.

This photovoltaic power plant has been designed under the consideration of using a technique for further projects in the future.

The Consortium

SIEMENS - PPC - VARTA

is entrusted by the EC to instal - in the framework of an international research and development program - a photovoltaic power plant with a rated peak power of 100 kW for supplying electrical energy into the existing network of the island.

Technical Data of the Plant

- Rated output (peak) of the array:	100 kW (AM 1.5/25 °C/ 1.000 W/m^2
- Rated voltage of the array:	160 V=
- Rated battery voltage:	250 V=
- Rated voltage of the PV power plant:	380 V, 3ph, 50 Hz
- Number of solar modules:	800
- Rated voltage of solar module:	8 V=
- Rated output of solar module:	125 W
- Dimensions of solar module:	appr. 1.5 x 1.0 x 0.08 m
- Weight of solar module:	appr. 27 kg
- Mounting angle:	appr. 35°
- Total area of the PV power plant:	appr. 7.500 m^2

1. INTRODUCTION

The photovoltaic power plant for the Greek island Kythnos has been installed in addition to the existing diesel power station and a wind converter plant being constructed in 1982.

Kythnos is one of the Cyclades Islands and is located approx. 60 km south of Athens. The area of the island is 163 km^2.

The photovoltaic power plant supplies energy parallel to the existing diesel power station and the wind converters into the 15 kV network of the island. The aim is saving of high costly diesel oil for the electricity supply of the island consumers on Kythnos.

The type of operation called as "fuel saver" is investigated in respect to reliability, economy and performance.

The reliability of the energy supply for the consumers of the island Kythnos is increased by the installation of the p.v. plant.

The operation of the plant is automatically performed by a superimposed control.

The total area of the plant is approx. 7500 m^2, the net area of solar modules is 1200 m^2 for the installed peak output power of 100 kW.

To feed the electrical energy from the solar generator to the grid a special system technology was developed.

The efficiency of the plant is as high as possible for economic reasons.

The Kythnos Photovoltaic Power Plant enables a clean energy supply by conversion of solar radiation directly into electricity. The p.v. plant is well suited to the environment and shows a new way of electricity production with the aim to reduce the dependancy on oil.

2. GENERAL TECHNICAL DESCRIPTION

800 solar modules forming a photovoltaic power station of 100 kW peak convert solar energy directly into electrical energy. This energy is either stored in a battery system or fed directly into the existing island grid stressing that the losses are kept to a minimum. The photovoltaic power station exists in its essential items of the following components:

- Solar generator
- DC/DC-converter
- DC/AC-inverter
- Storage battery system
- Superimposed control

Principle see basic circuit diagram picture 1 and layout picture 2.

Solar Generator

The solar generator has a total capacity of 100 kW peak and consists of 800 solar modules type SM 144-09 rated voltage 8 V, short-circuit current 16 A at standard conditions AM 1.5, E = 1000 W/m^2, cell temperature 25 $^{\circ}$C. 20 solar modules are connected in series and form an array with a voltage of 160 V. Each 2 arrays are wired to a terminal box and connected in parallel. A total of 40 arrays with 20 terminal boxes are distributed over the four terraces of the total plant. The smallest solar generator unit which can be switched is wired to the switchgear container. 5 double-arrays are connected in parallel to form a 25 kW unit, 4 of these units give the total of 100 kW.

DC/DC-Converter

4 DC/DC-converters transform the input voltage of 160 V to the battery system voltage of 250 V. These converters are regulated to the optimum working point of the solar cells (MPP) in order to exploit the cells at the most. As soon as the battery voltage reaches the final charging voltage the working point of the solar generator will be moved to the area of low load, so that the battery is loaded with constant voltage.

The 4 DC/DC-converters will be controlled by the superimposed control with shifted pulses in order to lower the residual ripple of the battery current and by this to enlarge the battery life time.

Storage Battery System

As storage battery system a battery especially developped for long term storage with low self discharge, long life-time and low requirement will be used. Rated voltage 250 V, nominal capacity 2400 Ah. For the storage battery the micro-processed control device LOGISTRONIC was developed making it possible to take into account the state of charge of the battery in the automatic monitoring system of the solar plant.

DC/AC-Inverters

By DC/AC-inverters the DC-energy will be converted to AC-energy. Each of the 3 inverters (capacity 50 kVA each) will be switched in dependence of the electrical power to be supplied to the system. This method of control enables losses to be limited to a minimum, especially when working in the partial load range.

The control of the DC/AC-inverters is designed for 4-quadrant operation, so that at a later stage by an overall superimposed control system for the island network an optimization of the power flow can be achieved.

Superimposed Control

By the superimposed control the automatic mode of operation of the plant is done. This control system receives information of all essential currents, voltages, state of battery charge and all essential signals of the plant.

DC/AC-Outgoing Bus

The DC/AC-outgoing bus consists of a three-phase system 3 x 380 V/Mp. The energy delivered by the photovoltaic power plant is fed via cable to a pole transformer and from there into the 15-kV-network of the island.

3. ARCHITECTURAL AND ENVIRONMENTAL ASPECTS

Kythnos Island is typically Aegean: Gentle slopes of the ground, absence of flat fields, low mountains, very scarce vegetation. A few buildings are scattered around having clear, cubic forms and are in most cases whitewashed.

The site selected is located near the diesel power station on a slope oriented to the south.

The basic aim of the architect and the contractors was to model the plant of the photovoltaic power plant in a way that will harmonize with the surrounding scene and making use of traditional materials and techniques where applicable. So dry stone walls where used built from local stone slabs in a very characteristical way, white, cubicle housing for the equipments etc.

The plot was divided in four terraces, decending east to west matching the contour of the land. The terraces are slightly inclined, so as to keep the height drop at the end of each terrace to one meter. Dry stone walls were built to retain the soil and to border the area.

The three double-containers with system equipment and batteries were white colored and positioned at the north side of the plot, where they cannot shaddow the solar modules and are easily accessible from the road.

As picture 3 shows the embedding in the landscape turned out well.

4. OVERVIEW OF NEW TECHNOLOGICAL DEVELOPMENTS

At the Kythnos solar power plant the new developed large area solar modules SM 144-09 are installed. These modules are built up of 144 monocristalline silicon solar cells each of 100 mm diameter.

For the conversion of the DC energy special transistorized DC/DC-converters and DC/AC-inverters were developed. Special emphasis has been given to high efficiencies, also at low loads, high reliability, operation under high temperatures etc.

The developed technology allows erection of solar plants for DC- and AC-supply within a high range of power.

5. MODULES

For the erection of the plant Siemens large area modules SM144-09 were used. With dimensions of appr. 1.5 x 1.5 x 0.08 m and a weight of appr. 27 kg they are still easy to handle. The installation of large area solar modules has the advantage of:

less installation work
less cabling connections
easier maintenance
simplier supporting structure

The total number of 800 modules are arranged in 4 array fields, each 25 kWp.

6. ARRAY AND CABLING

The total of 40 arrays each consisting of 20 modules connected in series are placed on four terraces. The mechanical arrays are identical with the electrical series connection of the modules. Two of this series connections are connected in parallel in a terminal box. When designing the cabling of the fields great importance was given to low cable losses, easy survey of the array field, easy maintenance and easy fault location.

The cabling is carried out with stranded single core cable with direct connection of the modules within an array and underground installation to the equipment container. The cabling between the 3 containers and to the pole transformer is also underground, the 15 kV connection to the grid as overhead line.

7. ARRAY SUPPORT STRUCTURE

The supporting structures are simply designed by using H-section beams IPBL 120 and tubes of 114 mm diameter for the upper mounting tube and 60 mm for the lower mounting tube. Picture 4 shows a supporting structure, picture 5 a part-field during erection. According to the geographical position of the site the modules are adjusted with an inclination of 35°. The distance between the arrays is optimized between shaddowing by the next array in wintertime and requirement of space for the total field.

This construction gives low costs for the erection of the array and placing of the modules, exact positioning of the modules, easy cleaning etc.

Picture 6 gives a view to a completed array.

8. LIGHTNING PROTECTION

The lightning protection system consists of two parts:

8.1 Mechanical Protection

By means of lightning protection rods the complete solar generator is protected against direct strokes into the generator (Ref. pict. 7). The protection space was designed according to VDE-standards.

8.2 Electronical Protection

Despite the mechanical protection there can arise overvoltages caused by the funnel shaped distribution of the ground potential around the rod that is hit by a lightning stroke (Ref. pict. 8).

To avoid uncontrolled sparkling there are lightning arrestors installed at both ends of the cabling between solar generator and DC/DC-converter. In this way transients from the DC-side are kept off from the power matching units. The blocking diodes of the subarrays are protected by varistors (Ref. pict. 9).

The AC-output of the plant is protected by conventional lightning arrestors. The lightning arrestors are choosen with an ignition voltage that limits the peak voltage to the voltage the insulation has to stand according to VDE-standards.

9. POWER CONDITIONING AND BATTERIES

9.1 Input from Solar-Field

The energy produced in the solar field is supplied in 20 strings - each 5 kWp - to the system equipment container. Each string can be switched in the DC/DC-converters. An indication of the circuit state is given in the mimic diagram.

9.2 DC/DC-Converter

Characteristic features

Maximum power extraction from the solar generators for each incident radiation power by means of maximum power point (MPP) regulation.

High efficiency due to electronic DC/DC-inverter principle.

Long battery service life due to IU (constant current/ constant voltage) charging method, low residual ripple.

Minimum maintenance

High degree of reliability

Automatic switching in and out of the DC/DC-inverters when a preset minimum incident radiation power is exceeded or not attained.

Reverse current inhibitor from the battery back to the solar generator.

High temperature range

MPP regulation

Due to the MPP regulation the power input of the DC/DC-inverter is always regulated such that the maximum power is extracted from the solar module groups. This lies at the bend of the IU characteristic of the solar module.

As these characteristics fluctuate greatly depending on incident radiation power and temperature, additional power is won by using MPP regulation.

9.3 Battery Bus

The DC-energy is collected and distributed at the DC-battery-bus. The 4 DC/DC-converters are feeding to the bus, the DC/AC-inverters take their energy from the bus and the battery takes or feeds the balance energy. All in- and outputs can be switched and are protected by fuses.

9.4 Battery and Control-Device LOGISTRONIC

The battery, which is installed in two 20 feet-double-containers has the following technical data:

Nominal capacity	2 x 1200 Ah (paralleled)	
Number of cells	250 in total (125 in series)	
Rated voltage	250 V dc	
Total stored capacity	600 kWh	
Dimensions per cell	length	307 mm
	width	383 mm
	total height	525 mm
Weight per cell (with acid)	134 kg	

The cells had been delivered to the site in dry pre-charged condition. After installing in the containers and filling with sulphuric acid 1.23 kg/l the batteries were loaded to full capacity by solar energy.

The state of the batteries capacity is observed by the micro-processed control device LOGISTRONIC. The charging and discharging currents are evaluated under consideration of the loading factor, battery characteristic, battery temperature etc. The actual battery capacity is indicated and also given in digital values to the superimposed control.

9.5 DC/AC-Inverter

Characteristic features

High efficiency due to transistorized inverters operating on the pulsed principle.

Self-commutated inverter resulting in no-loading of the supplying system with inductive reactive power.

Possibility of installing independent system supplies.

Possibility of operating several inverters in parallel.

Uniform distribution of active and reactive power during parallel operation.

Inverters switched on and off depending on the power requirement.

High impulse load capacity.

Rapid switching in possibilitiy for covering sudden load requirements when operating without other energy sources in parallel.

Minimum maintenance
High degree of reliability
High temperature range

9.6 Superimposed Control

The superimposed control takes information from all parts of the plant (Ref. pict. 1) and optimizes the energy flow within the system of solar-generator, DC/DC-converters, DC/AC-inverters, battery and load and the set value, given from the diesel power plant.

For the future an overall load management optimizing power-sharing between solar-, wind- and diesel-generators should be considered.

9.7 AC-Bus

The 3 DC/AC-inverters are feeding in parallel to the AC-bus. Depending on the actual load 1, 2 or 3 inverters

are switched on by the superimposed control. When an inverter gets the signal to start it automatically runs up, synchronizes, hooks up to the line and takes over its share of the load. Following is an acknowledge to the superimposed control.
The AC-Energy is fed via a transformer to the 15-kV-grid of the island.

10. OPERATIONAL CONTROL AND MONITORING SYSTEM

The solar power plant operates automatically by means of a superimposed control.

The essential operating datas are recorded by line recorders. In addition each converter and inverter as well as the battery have measuring instruments.

The operation, circuit state and faults of the different parts of the power plant can be supervised on a mimic diagram.

From this mimic diagram the DC/DC-converters and the DC/AC-Inverters can be switched on and off electronically.

In the case of a fault an alarm is given to the diesel power station and is indicated on the alarm annunciator combination.

Details see picture 10 and 11.

11. SYSTEM INTEGRATION AND INSTALLATION

The system is installed in a 20 feet double container. The cubicles are arranged in a U-form, which gives a good survey of the units - see picture 12 and picture 13.

To carry off the heat arising by the power losses of the converters etc. a ventilation for the equipment container is installed. To save energy the ventilation rate is controlled in dependance of the inside temperature of the container in a wide range. At low temperatures the ventilation is cut off.

The batteries are in installed in two 20 feet-double containers.

The output of the solar power plant feeds the 15-kV-grid of the island by a transformer. With the self-commutated inverters for four-quadrant operation the solar power plant gives practically no additional harmonics and inductive load to the network.

The four-quadrant operation allows the plant to supply and receive inductive, resistive and capacitive power.

This enables to supply either the received solar energy directly resp. after storage in the battery into the island grid or to receive abundant energy for example from the wind generators from the grid and to store it in the battery.

This energy will than be fed at a later time into the net. For this purpose the solar power plant can be controlled by presetting a setpoint value.

When widening the island network by a overall superimposed control the interconnected system operation between the three power generating systems diesel/wind/solar could be optimized.

12. CONCLUSIONS AND FIRST RESULTS

The solar power plant has been installed and put into operation in time. Until the deadline end of June 5.500 kWh has been supplied into the island's grid.

The experience verifies, that the arrangement of the plant gives a good clearness to the operator. The measuring and recording instruments allow a good survey to the actual and past operating conditions and the mimic diagram shows the circuit state and eventually faults very clear.

The practical operation showed that the new developed DC/DC-converters with MPP-control allows an optimal utilization of the solar radiation at the hard differing temperatures of the modules in connection with very good operating conditions for the battery.

The new developed DC/AC-inverters meets the expectations of high efficiency, quick control, good accommodation to variable load and frequency etc. The good experiences of the short operating period have to be proved at the further time.

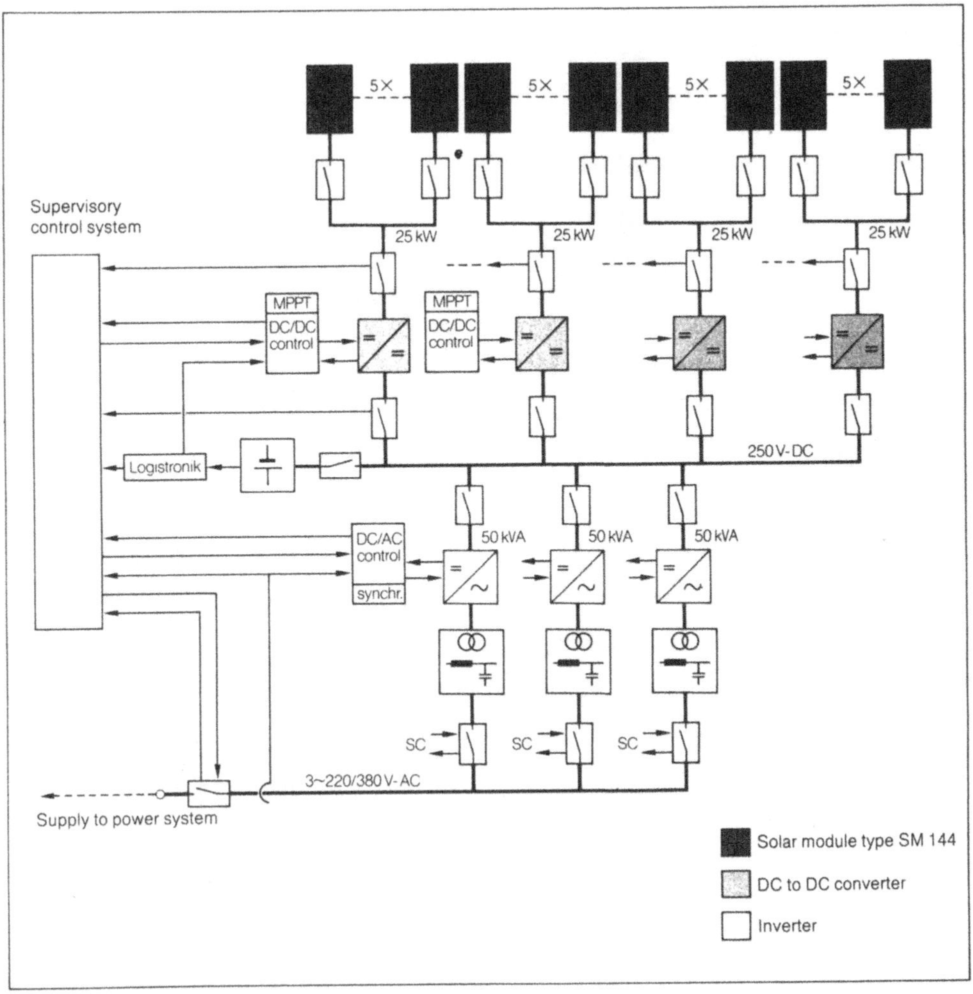

Picture 1 - Basic circuit diagram

Pict. 2   Layout

Pict. 3 - Embedding of the plant in the landscape

Pict. 4 - Supporting structure during erection

Pict. 5   Solar part field during erection

Pict. 6   Completed array

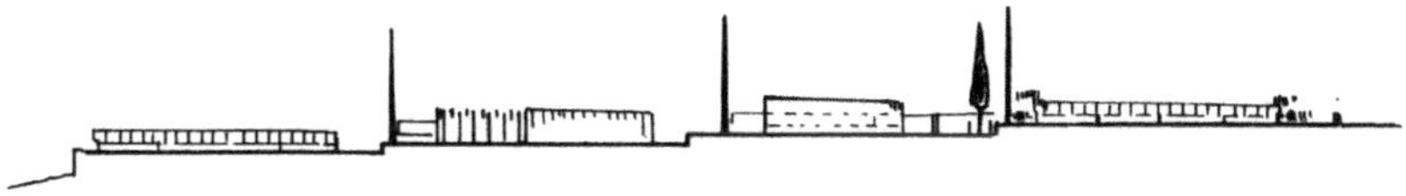

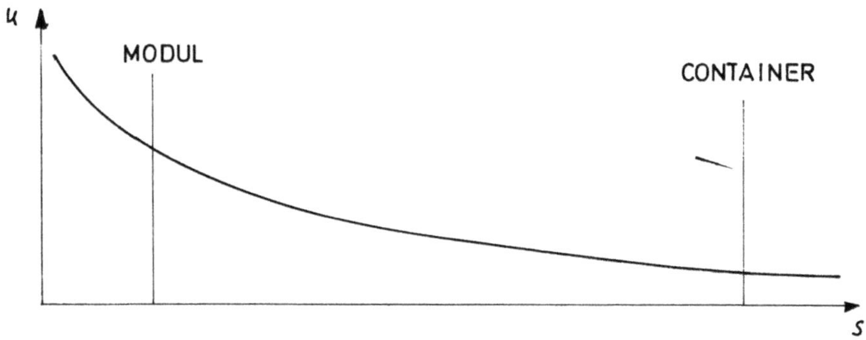

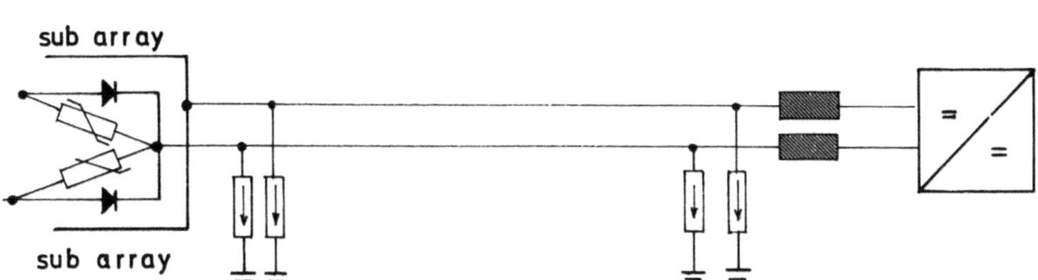

Pict. 7...9 Lightning protection

Design of

Operational Control Panel

The 2-Panel Control Cubicle Comprises:

I. On the Frontside

1. Mimic diagram

2. Two colour line recorder for:
   Current of DC/DC-converter No. 1 and No. 2

3. Two colour line recorder for:
   Current of DC/DC-converter No. 3 and No. 4

4. Three colour line recorder for:

   - Battery current
   - Battery voltage
   - Battery capacity

5. Three colour line recorder for:

   - Total active power supplied
   - Solar irradiation
   - Ambient temperature

6. Alarm announciating combination with optical and acoustical signalisation for 12 signals:

   - DC/DC-converter No. 1: Interruption
   - DC/DC-converter No. 2: Interruption
   - DC/DC-converter No. 3: Interruption
   - DC/DC-converter No. 4: Interruption
   - DC/AC-inverter No. 1: Interruption
   - DC/AC-inverter No. 2: Interruption
   - DC/AC-inverter No. 3: Interruption

   Plus

   - 5 additional signals (as spare) for example for battery-supervision/earth fault detection

7. Mode of operation selector switch "Hand-/Automatic Operated"

II. Inside the 2 Panel Control Cubicle

- The superimposed control
- The LOGISTRONIC
- kWh-meter
- Battery earth fault supervision
- 24 V power supply
- Flashing relays for the lamps on the mimic diagram

etc.

Picture 10 - Control panel

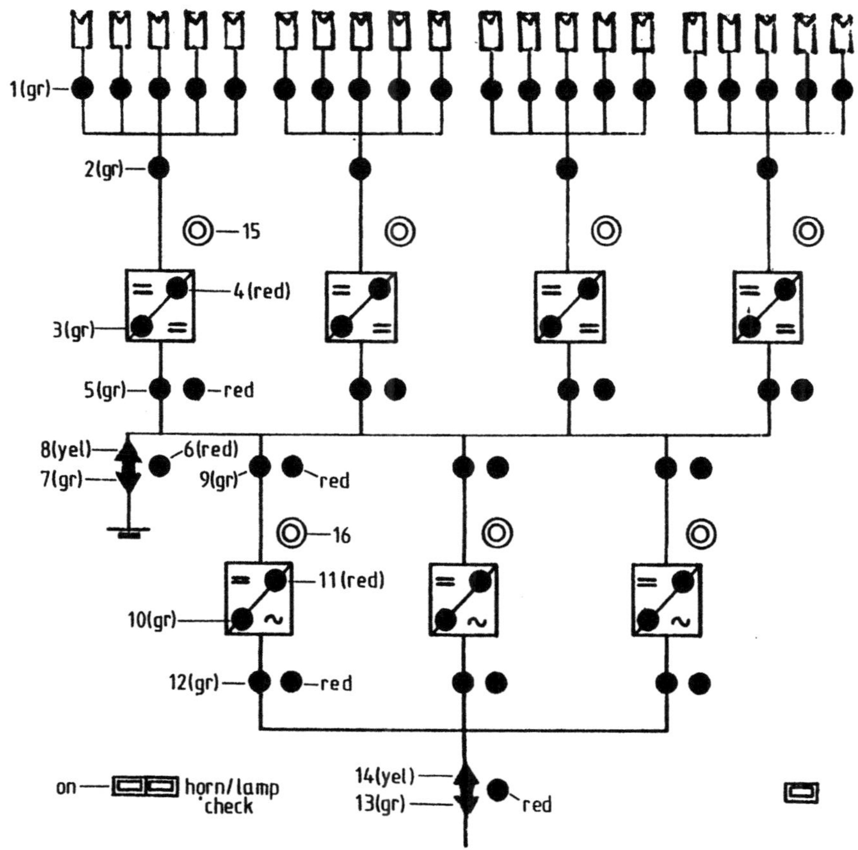

Mimic Diagram - Signal Lamps

1. Incoming switch from solar modules
2. Incoming switch from DC/DC-converter
3. Operation: DC/DC-converter
4. Interruption: DC/DC-converter
5. Outgoing switch of DC/DC-converter
6. Interruption: Battery
7. Operation: Battery - charge
8. Operation: Battery - discharge
9. Incoming switch: DC/AC-inverter
10. Operation: DC/AC-inverter
11. Interruption: DC/AC-inverter
12. Outgoing switch: DC/AC-inverter

Outgoing switch to diesel power station:

13. - feeding
14. - receiving

Push-to-Lock Key For electronic switch-operation:

15. for DC/DC-converter
16. for DC/AC-converter

Picture 11 - Mimic diagram

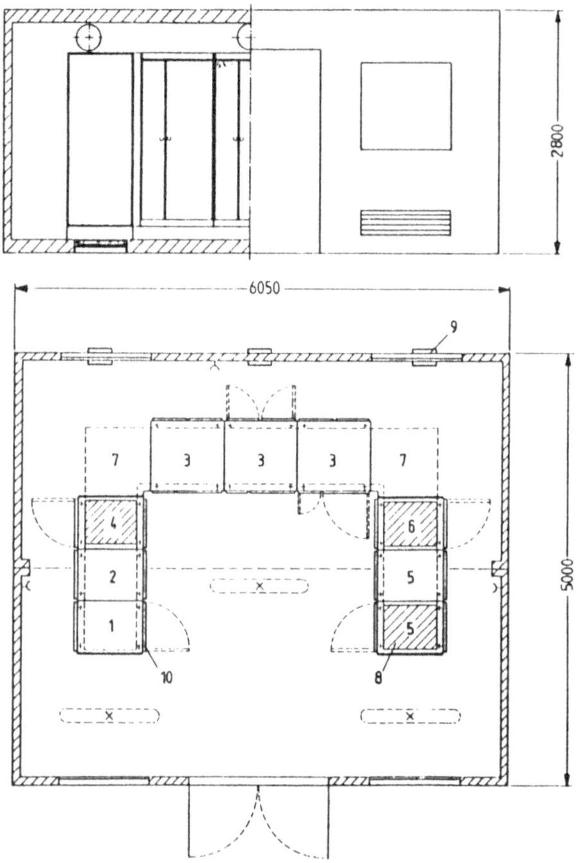

1 Cubicle instrumentation and control I
2 Cubicle instrumentation and control II
3 Cubicle inverter
4 Cubicle system connection
5 Cubicle DC-chopper
6 Cubicle battery bus
7 Duct for pipes
8 Cable entry container
9 Fan
10 Mimic diagram

Equipment installed	5,5 t
20 Persons	1,5 t
Container	6,0 t
Total weight	14,0 t

Picture 12 - Housing for switchgear

Picture 13 - Converters, inverters, switchgear, control

TREMITI DESALINATION PLANT

Location, country	: Tremiti Islands, Italy
Rated peack power of array	: 65 kW
Contract number	: ESC-P-064-I(S)
Duration of contract	: 37 Months,1 March 1981- 15 March 1984
Total budget of project	: 2.672.000 ECU, CEC contribution 455.000 ECU, 17%
Name of other funding bodies and their shares of total cost	: Cassa per il Mezzogiorno 2.217.000 83%
Contractor	: Adriatica Componenti Elettronici S.p.A. Ansaldo S.p.A. Indelet Industriale Elettrica S.p.A. Italenergie S.r.l. Petrochemical Intern.Instr. Co. S.p.A. Tecmar S.r.l.
Project leader	: Mr. Fulvio Fonzi c/o Italenergie I-67039 Sulmona Tx(0043) 601105; ph(0039864) 52934

Summary

The Tremiti project will assure the drinkable water supply to the Island of S. Nicola in the Adriatic Sea. The system is a stand-alone plant for sea water desalination using the reverse osmosis process, energized by a photovoltaic power plant. A p.v. array will give a 65 kWp nominal peak power. MPPT/MBC converter will feed the main d.c. bus and the storage battery which assures the uninterrupted operation of the desalination plant during overcast periods. The total plant area is 4500 sq.m; 1800 of them covered by the p.v. field: the layout of the rows of modules has been designed to optimize the efficiency of the array subsystem, to minimize atmospheric overvoltage, and to assure easy operation and maintenance. Good fitting with the natural characteristics of land is obtained. The final solution of the supporting structure satisfies to the last Italian codes for seimic areas: pv modules are bolted to galvanized steel beams supported by walls of prefabricated vibrated concrete blocks. The electric appliance load will consist in a.c. induction motors driven by single inverters. The R.O. plant will produce about 5000 cu. m. of water with an average power requirement of 13 kWh/cu.m. A power flow supervisory system will optimize, in every possible operating condition and system status, the utilization of available energy. A data acquisition system will monitor continuously the system and record the relevant parameters.

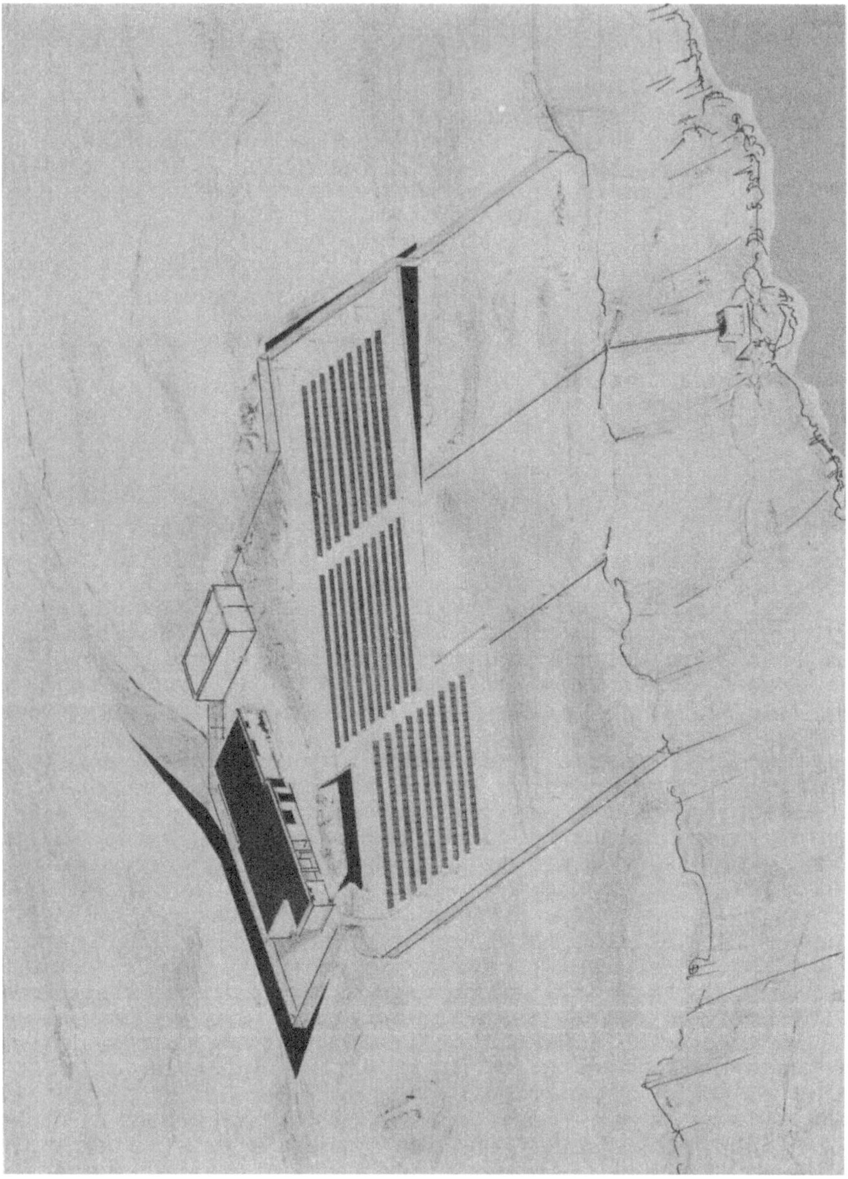

Introduction

The Tremiti project is aimed to energize a sea water desalination plant to assure the drinkable water supply to the island of S.Nicola in the Adriatic Sea.
The Tremiti Islands have no natural sources and the water supply for the few tens of residents and the nearly two thousand summer holyday tourists is now assured by tankers sailing from the continent. The desalination process chosen is the reverse osmosis (R.O.) because of its high efficiency (less then 13 kWh/m^3 of drinkable water) and of its simple operation, allowing an easy automatic management. The size of the desalination plant was fixed in a potential output of 30 m^3/day, based on a 24 hours/day operation. The yearly production required for the S. Nicola Is. amounts to 5000 tons approximately, corresponding to nearly 65000 kWh/year to be supplied by the photovoltaic plant, no grid being available on the site. The autonomous operation of the desalination plant all the year round, excepting two or three stops during the longest overcast periods in winter may be assured by a peak power of the p.v. array of 65 kW. This figure was fixed taking into account the system losses, mainly in the energy storage sub-system, the auxiliary loads, and the expected average hours of insolation on the site. The whole plant will occupy a 4500 m^2 area on the southern coast of the S.Nicola Island between the ridge of a 44 m high cliff and the crest of the hill. The p.v. array covers 1800 m^2: the building housing the remaining equipment of the p.v. power plant and the R.O. plant is erected along the northern border of the area and is less than 3 m higher than the surrounding ground. The natural slope of the terrain and the masking assured by the cliff have allowed a very compact layout, giving only minor disturbance to the landscape of this uninhabited and scarcely frequented area of the Island.

The block diagram of the power plant was finally defined through an accurate analysis of the load power requirements and the comparison of the efficiency of the different technical solutions. The p.v. array, splitted in three partial arrays, is connected to the general d.c. bus through three identical d.c. to d.c. up-converters, controlled in order to perform the MPPT/ MBC functions. The storage lead-acid battery is connected to the d.c. bus, and when completely charged may assure 4 to 5 days of operation of the desalination plant, in the absence of p.v. array output. The main loads are pumps, operating at constant pressure and flow: the electric drive solution finally adopted consists in induction a.c. motors fed by d.c. to a.c. inverters. During the start of the motors the output voltage and frequency of the inverters are increased proportionally, getting costant torque and current conditions: thus no oversizing of the inverters is required to support high starting currents. As shown in Fig.1.1 the power plant is a fully d.c. system and each a.c. load is equipped with its own inverter. This solution compaired with an a.c. power system requires a higher number of inverters of smaller size but assures a better efficiency and improves the continuity of operation because all the inverters work at their rated power output and at the some time few stand-by units are

FIG. 1.1 POWER PLANT BLOCK DIAGRAM

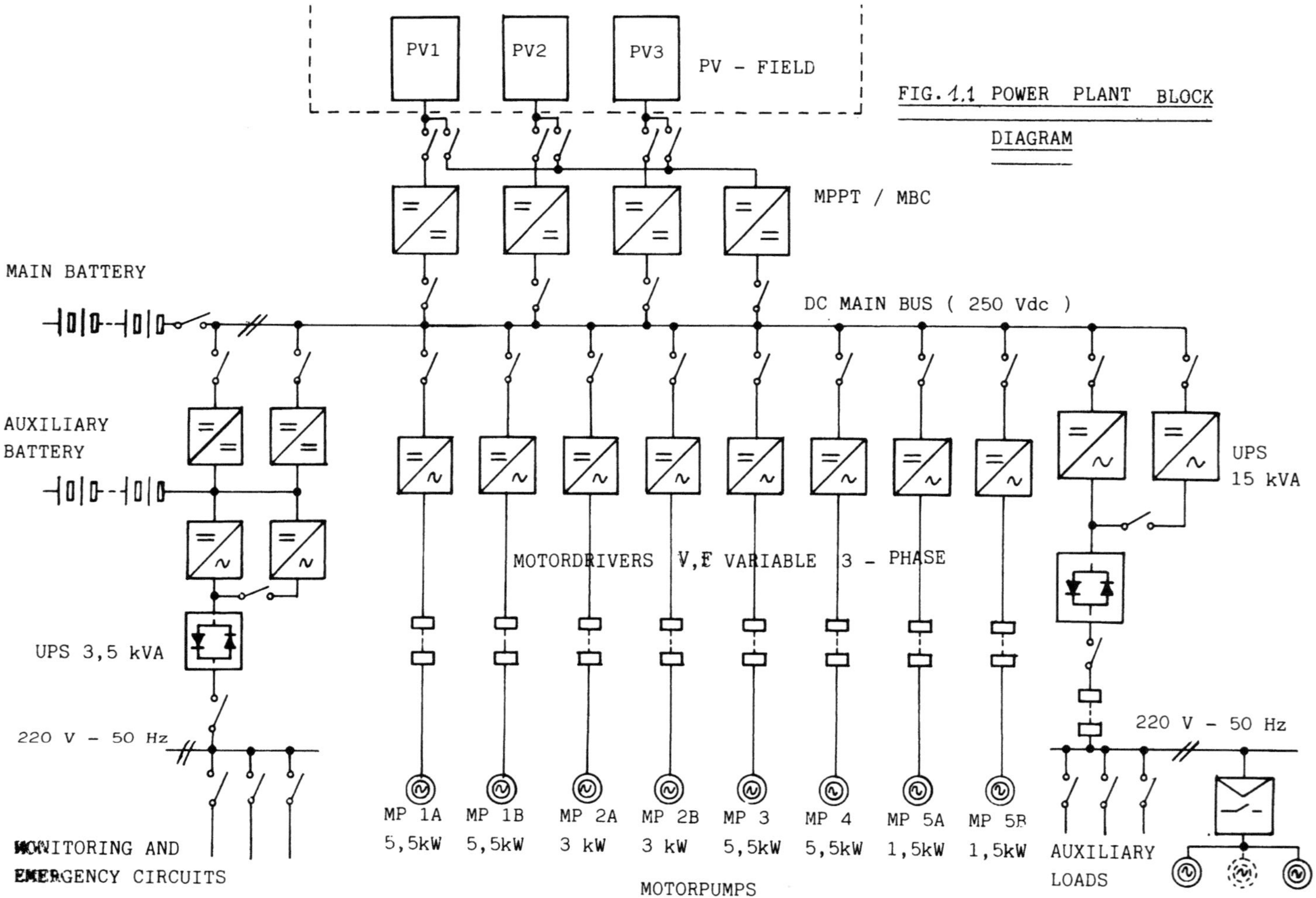

required for continuous operation. An additional ac single phase inverter supplies auxiliary loads of small motors, lighting and fans. An UPS, including a chopper charging a floating lead acid battery and a single-phase inverter, assures the supply to the power management and monitoring equipment even when no power is available from the pv plant.

Location: S.Nicola Island, latitude 42°07 N-longitude 15°32' E. (Tremiti islands in the Adriatic Sea, 24 n.m. from the continent shore)

Site: South-wards sloping (11°) ground, average height 58 m a.s.l. Fractured rocks with low vegetation low seismic Lightning density less that 3 strokes/km^2 x year. Temperature range -5°C to +40°C, RH yearly average 72%; rain 440 mm/year; snow fall less that 1 day/year.

Estimation of solar radiation: no solar radiation data are available for Tremiti Islands, thus interpolation among data sets recorded by the Aeronautica Militare Network had to be carried out. Since there is no large difference between the total solar energy collected by a fixed tilted surface and the maximum energy amount available if a one-dimensional tracking system is used, a fixed tilt angle of 22° has been chosen with a loss in energy collection smaller than 5%. The spacing between rows, taking into account the mean slope of the terrain (11°), will be 1,5 m c.a. to avoid shadowing even in the most unfavourable condition (21 Dec. at sunrise and sunset).

Application: sea water desalination plant, using the reverse osmosis process (R.O.); the produced drinkable water will be let in the Island acqueduct reservoir. The nominal production rate of the plant is 30 m^3/day. The total production capability is splitted in 2 sections that may be operated independently.

The rated power of the R.O. plant is 16 kW. The total installed power amounts to 42 kW (including stand-by). The loads are induction motors driving pumps; in normal operating conditions they will work at their nominal power. The motors in the plant, as well as the motors of the sea water suction pumps, are fed by inverters whose output frequency will be 50 Hz during the normal run, and will increase from 5 to 50 Hz during the starting phase of each motor. The a.c. output is a stepped rectangular wave: the harmonic content is in the order of 30%; no filtering is requested provided the motor can withstand losses increased by less that 10% with respect to sinusoidal input. The remaining motors are in the range of some tenths of kW, they will be fed from a 50 Hz 220 V single-phase bus and will not require any particular starting provision.

The load diagram of the R.O. plant will fit the available energy diagram, in order to assure, on the yearly basis, the optimum efficiency of the whole system, the difference between drinkable water production and consumption will be stored in, or taken from, the acqueduct reservoir, as shown in Fig. 1-2.

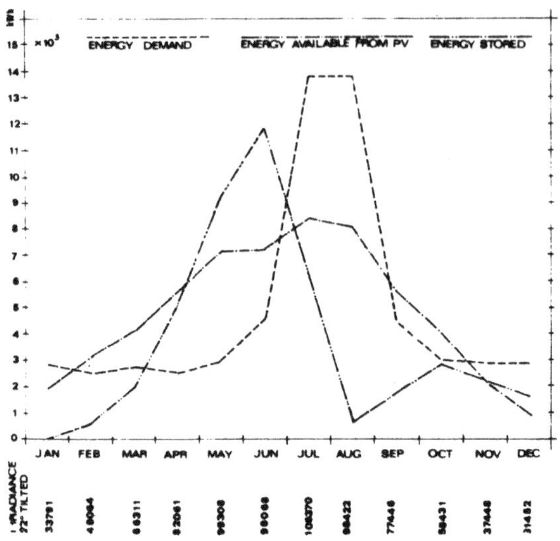

FIG.1.2 : Energy diagram

2. <u>General technical description</u>

The pv plant consists in a field splitted in three sections which can be operated indipendently and in a dc/dc conversion system including three converters performing as MPPT/MBC. The converters feed a general dc bus to which is connected the main lead acid battery; from the bus are derived the supply lines of all loads. These are eight 3-phase induction motors with power ranging between 1.5 and 5.5 kW. Each one is driven by a square wave inverter, whose output voltage and frequency are controlled during the starting phase of the motor in such a way that the starting currents do not exceed the nominal rate of the inverters while steady state voltage and frequency are kept constant.

The power plant has been designed to allow two operational modes:

- supply of loads from pv field and/or from battery
- battery charging with or without feeding of loads.

Most effective use of the available power is thus obtained in every situation.

Auxiliary loads, chemicals injectors, lighting and fans are supplied by a single-phase inverter while the monitoring and power management system is connected to a UPS unit which includes a chopper, lead acid battery and single-phase inverter. Furthermore the same UPS energizes all the starting switches of the pv plant. The monitoring subsystem includes transducers for the measurements of all physical quantities useful for both process supervision and off line analysis of the overall operational performance. Acquisition and processing of data are performed by means of a central unit which includes two dedicated micro-processors; row and processed data are hence displayed on a CRT colour monitor. Data for performance analysis are stored on cassette for further processing. Each section of the plant is equipped with own alarm and automatic block devices which send to the central unit signals to be displayed on a board. The central computer issues diagnostic messages to suggest to the operator proper corrective procedures. Automatic operation of the whole system will be possible under the supervision of the central unit, by means of a management program which shall be developed on the basis of a significant set of experimental data.

3. Link application system/generator

The performance of a stand-alone system energized by solar power can be optimized through an accurate matching of the peak power of the solar generator, including power conditioning equipment and of the appliances loads and through an appropriate definition of the storage capacity. The hardware cost and thus the yearly amortization rate are primarily determined by these parameters. The final solution adopted for the p.v. power plant, as well for the elctric power requirement of the R.O. plant allows to achieve the objectives of:

- high efficiency of the energy transfer from generator to load;
- elimination of all unnecessary power conditioning and of the relating components.

As compared with previously considered solutions the last one delivers more energy at lower cost.

The production volume will depend on the power management strategy, whose goal might be defined as the utilization at the highest possible effectiveness of instanteneous generated power.

The Tremiti project was conceived on a basis allowing a high degree of flexibility in the power dispatching, assuring the most effective use of power available from the p.v. array. The desalination plant consists in separate sections that are normally operated simultaneously for the highest possible process efficiency. When the array output exceeds the maximum power requirement of the R.O. plant, two different energy storage systems are available: i.e. the lead-acid battery and the sea-water reservoir. Of course the two energy accumulations must be handled coherently in order to assure the maximum of autonomy to the R.O. plant. The principles of the power management strategy have been defined according to the priorities listed below, (fig. 3-1);

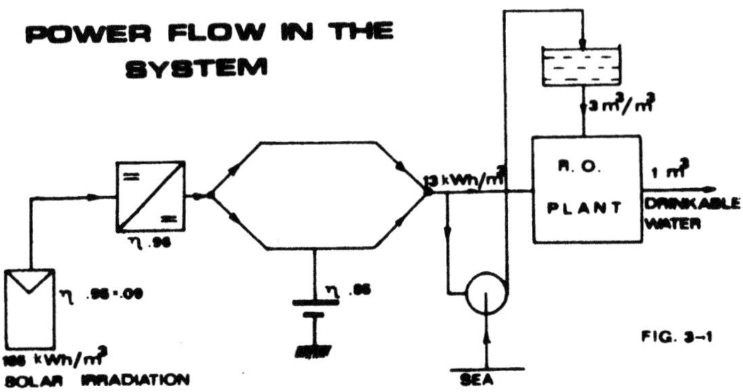

FIG. 3-1

a) on a long term basis: electric energy and sea water stored must be kept at balanced levels, sufficient to assure the R.O. plant operation at a daily production rate determined on the basis of the most likely seasonal sequence of overcast days;
b) on a daily basis: during sunshine, the R.O. load will be the highest compatible with the replenishment of storage. If storage levels satisfy to condition a), the R.O. load must not require discharge of the battery and sea water reservoir in the average day.
The strategy will result in:
- minimizing the stops of R.O. plant, due to lack of electric energy and pretreated water; these stops will be limited to exceptionally long periods of bad weather;
- maximizing the operation time of R.O. plant under direct supply from p.v. array; thus excluding the energy loss in the storage system;
- optimizing the ratio: output (m^3 of water) to input (solar energy) on annual basis.

This strategy has been checked, using a simulation program specifically developed, and, as input to simulation, historical series of insolation, available for the Pescara met station (A 8 day sample run is shown in fig.3-2).

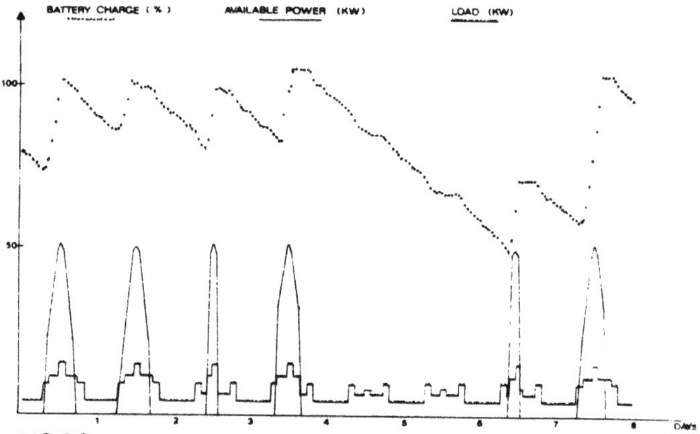

FIG.3.2 8-days run's energy balance

4. <u>Architectural and environmental aspects</u>

An accurate survey on the site has suggested to modify the geometrical configuration of the p.v. field from the previous rectangular shape. The new lay-out fits better with the natural slope of the rocky ground, and reduces the impact on the landscape of the array surfaces, improving the effects of the natural masking provided by the high coastal cliff. A further advantage is the reduction of escavations an liveling required.

The inclusion of Tremiti in the moderate seismic risk areas has required a revision of the structural design of the building to comply with the Italian codes. A completely new design of the building has been developed, with the final result of a more rational internal space distribution and of a more attractive external appearance. The limited height of the building and an appropriate external painting will minimize the interference with the natural features of the surroundings. A moderate height vegetation will be implanted around the building and the sea water reservoir, to improve the integration of the plant in the environment.

5. Overview of new technological developments.

P.v. modules of both manufactures have incorporated all the technological improvements achieved since the project start. Their actual performances are better than those assumed at the beginning of the project as shown by the table of Chapter 5.

Two aspects of the power plant design may be considered as technological advances in this area, namely:

- the original solution adopted consisting in a completely d.c. power generation plant feeding a.c. application loads, using the latest technological advances in the area of electronic motor drives offers remarkable advantages with respect to both solutions: (i) an a.c. power plant with large inverters, requiring oversizing and working for long periods well below their rated power, with a reduced efficiency; or: (ii) a d.c. power plant and application using d.c. motors having lower efficiency and reliability;
- the original power management strategy developed results from an objective analysis of the problems of a stand-alone system, more than from the extrapolation of solutions usually adopted in conventional supply systems for which there are no constraints on energy availability.

6. Modules

Two types of modules will be employed:

The Siemens produced module SM 144-09 including 144 monocrystalline silicon cells and two by-pass diodes;

the Ansaldo produced module AP 35 including 36 semi-crystalline silicon cells and two by-pass diodes.

The main typical characteristics of the modules are listed below:

	Siemens	Ansaldo
V_{oc}	10,6 V	6.66 V
I_{sc}	18 A	7.5 A
W_p	130 W	35 W
Insulation	2000 V	2000 V
Weight	27 Kg	10.2 Kg
Dimensions	1470x1020 mm	1298x343 mm
Frame	Al profile	Al profile

7. Array and cabling

The modules are connected in elementary series (21 Siemens modules, 33 Ansaldo modules) each one constituting a sectionable sub-array giving in AM1 conditions an operating V=160 V. The outputs of each string of one partial array are parallel connected in a general board inside the plant building. For each string a two-wire armored cable (2x10 mm^2) starting from a junction box and running in underground tunnels to the building is provided. Three partial arrays will thus be formed: two, each one connecting 9 Siemens strings, the third one 19 Ansaldo strings. The lines are protected against induced over-voltages by insulated Cu wires running immediately below the modules, that assure an efficient screening against induced voltages, the frame of each module being frankly earthed. Fast connection by means of plugs-receptacle is provided for Ansaldo modules. The layout of sub-array in the field (Fig. 7-1) minimizes the wiring losses, and the areas of parasitic loops. The expected mismatch loss is less than 3%, the wiring loss is also less than 2%.

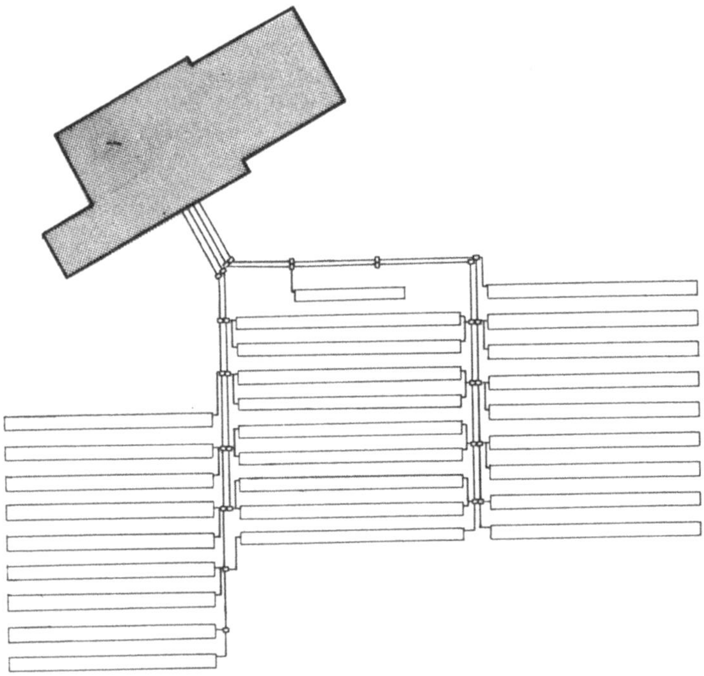

Fig. 7.1 ARRAY FIELD AND WIRING

8. Array support structures

The design of array support structures has been modified, in what concern the shape and the materials of the supports. The decision of the change was imposed by the recent classification of the Tremiti Islands in the second class of seismic risk (moderate risk). The loads analysis previously limited essentially to the forces resulting from the winds pressures on the modules (see Report in the proceeding of the 30 Nov.-1 Dec. 81 meeting in Bruxelles) was revised to take into account the forces deriving from the seismic accelerations, acting along different axes. A new configuration of the single support was chosen, i.e. vertical walls of vibrated concrete precasted blocks whose internal holes are filled with reinforced concrete tied to the linear foundation concrete beam. The fixation of modules was also revised in order to simplify and speed up the mounting operation of modules to the galvanized steel beams. The clear spans between supports in the different zones of the p.v. field are kept identical to those of the previous design. A pictorial view of the new solution, with some mechanical details are shown in Fig.8-1. A moderate increase in overall cost of the support structure has been obtained against a significant increase in mechanical characteristics. The type of masonry adopted matches quite satisfactorily with the features of local architecture.

9. Lightning protection

The probability of direct strokes on the p.v. array is very low (in the order of .5% year). Therefore no protection against direct strokes has been foreseen, the inconvenients and risks deriving from the presence of poles and/or guard wires in the p.v. field having been estimated too penalizing.
The protection against induced overvoltages is assured by the proper grounding of all structures, including module frames, by minimizing of the area of all loops of the wiring and by the use of metal oxide dischargers.

10. Power conditioning and batteries

The p.v. power plant, in its final design, is essentially a d.c. power generator, whose output bus have a nominal voltage of 250 V. Additional outputs are:

- a single phase 220 V, 50 Hz, 15 kVA for the auxiliary pumps of the R.O. plant, lighting, fans.
- a 1 phase 220 V; 50 Hz; 1 KVA assuring the supply to the auxiliary equipments monitoring, starting switches of the power plant itself, emergency lighting.

The d.c. power supplied by the 3 sections of the p.v. array at a nominal voltage of 160 V is converted to the bus nominal voltage of 250 V of the main bus and of the storage battery by means of three identical d.c. to d.c. static up-converters, using well proven semiconductors techniques. The rated power of each converter is 25 kW. The up-converters will be automatically controlled so that they will perform the MPPT/MBC functions, assuring the optimum load to the PV array in every insolation condition, and the appropriate output voltage to the main bus. The splitting in three sections of the primary generator (PV array converter) improves the operational availibility of the system, and lowers the maintenance costs. The main bus voltage will float around the nominal value of 250 V, depending on the state of charge of the storage battery connected to this bus, and on the direction of the actual power flow between the battery and the bus. The voltage will be contained between 290 V (battery being charged, having reached nearly full capacity) and 230 V (battery discharging at 20% of its capacity). These limits are compatible with the load (R.O. plant) requirements. The auxiliary ac. outputs will be assured by a conventional single phase inverter, and by a continuity static group consisting in a controlled chopper reducing bus voltage to 220 V. a floating lead-acid battery and a 1-phase conventional inverter.

The supervisory power management system will assure the most appropriate power flow, in the system, according to the

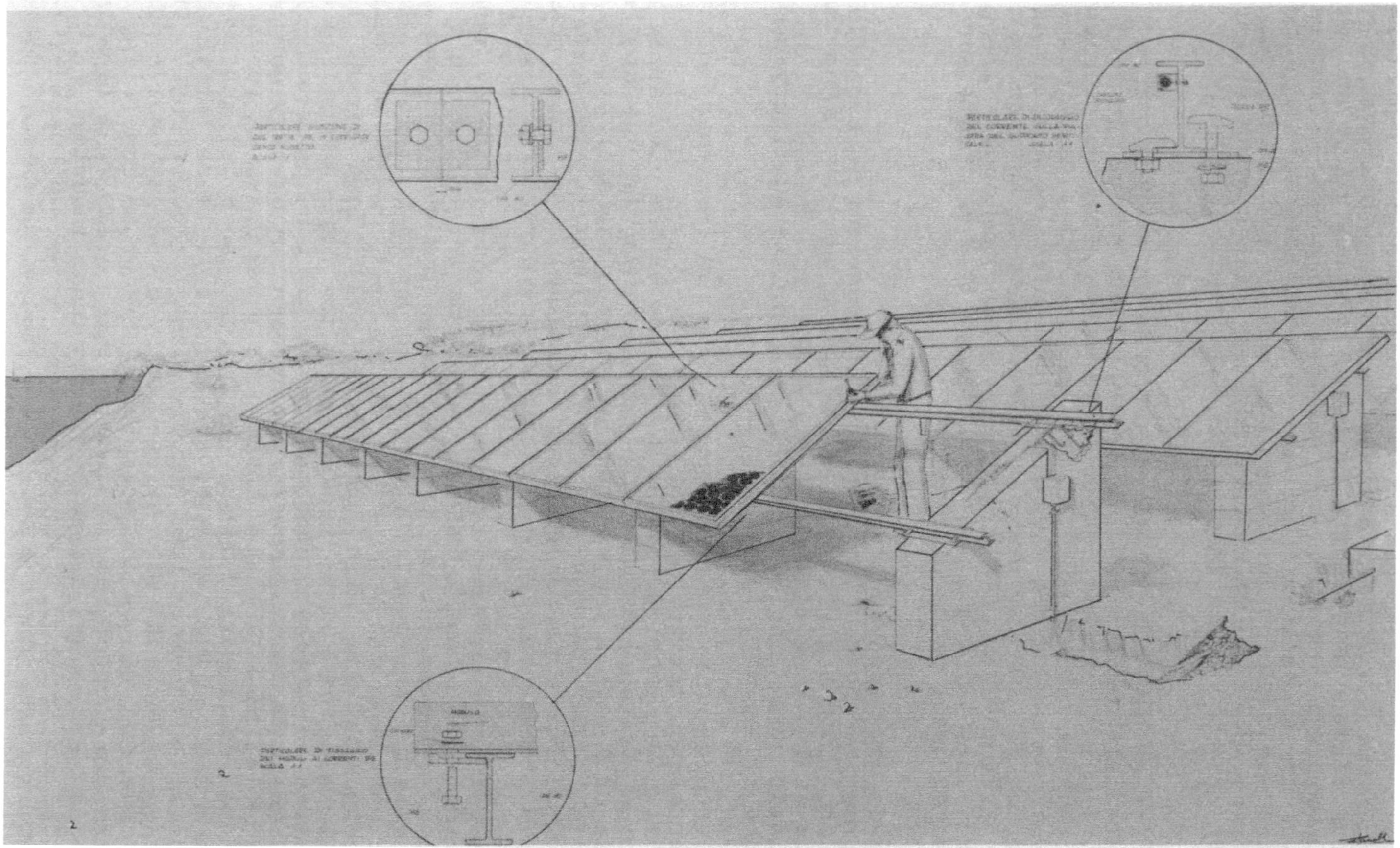

Fig.8-1 Array support structure

following synthetic table.

Array output	Battery charge	Power to load	Losses
yes	100%	full array power	generator only
yes	100%	array-brt. charge	generator
no	20%	from btr.rated for expected overcast duration	generator+ storage
no	20%	----	----

In a first running-in phase of system operation the supervisory system will process data and issue messages suggesting to the operator the most suitable configuration of the genrator/storage/application system. When a sufficient experience of this manual operation will be reached, automatic operation will be possible, by actuating switches and valves directly from the supervisory system outputs.
The main energy storage of the P.V. power plant is assured by a lead-acid stationary battery, having a nominal voltage of 250 V and a nominal capacity, at the discharge rate of 10 hours, of 2000 Ah. The battery will be made by a series of 125 cells of the Varta-bloc type, specifically developed for application to solar or wind power generators. They are characterized by a remarkable increase of capacity above the nominal value for low rate discharge. Thus for a discharge in 48 hours (yearly average of seasonal expected overcast durations) the available discharge capacity rises to 2700 Ah. The system energy storage capability is increased by a suitable dimensioning of the sea pre-treated water reservoir. Extra accumulation in this reservoir may be foreseen in case of a power output from arrays exceeding maximum R.O. plant power, and of a fully charged battery. The power flow supervisory system will assure in this way the integral utilization of incident solar energy.
The additional lead-acid battery for the auxiliary power supply feeding the monitoring and supervisory system will be of the same type but at a nominal voltage of 220 V and with a capacity of 250 Ah_{10}.

11. Monitoring system

The global system consists of 4 section:
- transducers section;
- actuators section;
- data gathering and handling subsystem;
- plant management subsystem.

The transducers section includes all sensor for the different variables, signal conditioning devices, interconnection wiring.

The actuator sector, in its first configuration is limited

to enabling switches. The data gathering and handling subsystem is fully controlled by a dedicated microprocessor; and includes, besides the CPU, all analog and digital I/O interface, RAMs memories, the specific firmware on EPROMs, peripherals for data display, alarms, synoptic visualisation, operator's I/O peripherals, mass storage (Fig.11-1).

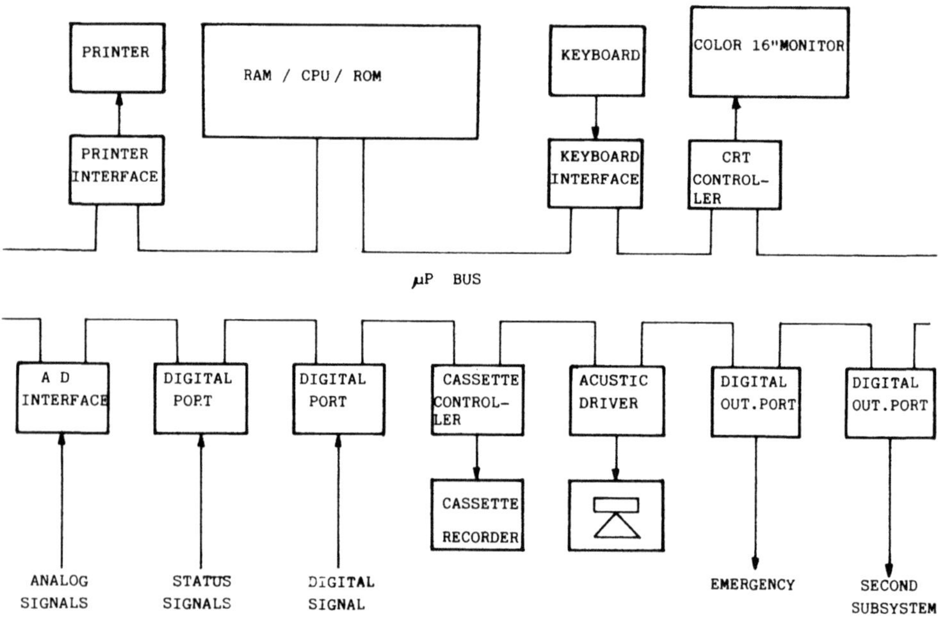

FIG.11.1 MONITORING SYSTEM BLOCK DIAGRAM

The management subsystem consists of a second identical microcomputer, developed for future on-line operation, with I/O peripherals, mass storage, and a firmware allowing operation through the use of a high-level language (Fig.11-2).
A panoramic bureau will house of all the hardware and the control desk.

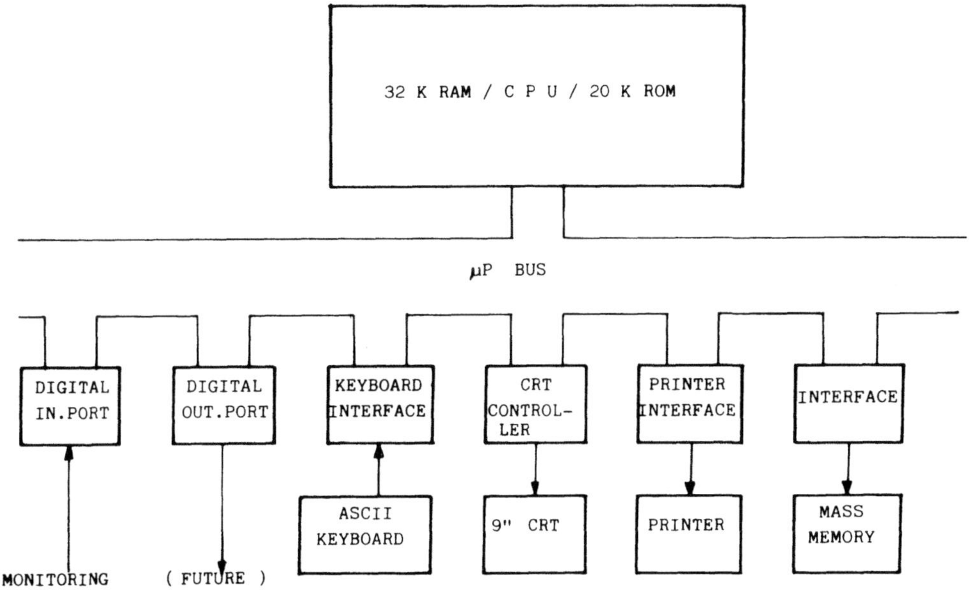

FIG.11.2 MANAGEMENT SYSTEM BLOCK DIAGRAM

The design criteria were chosen taking in mind the purpose of the pilot p.v. plant project, i.e. to acquire continuously and record accurate information about all significant functional and operational variables of the plant. Therefore, the assumed specifications for the monitoring system exceed the usual requirement, as to number of monitored variables, equipment reliability and availability.

Additional design criteria were:
- possibility of operation by unskilled personnel;
- lowest possible power consumption;
- high reliability in the unfavourable marine environment (100% RH, condensating, salt particles and chemical vapours contained in the air).

All sensors used in the monitoring system, excepted the insolation measuring devices, are active transducers giving a d.c. current output, function of the value of the measured variable. Therefore, no correction is required for the wiring resistance, and its temperature-variation.

The actuator section is limited to 8, expandable to 16 actuators for control of the plant.
The actuators are driven by commands issued by the microprocessor controlling the monitoring subsystem through appropriate interfaces.

The data acquisition and handling sub-system is controlled by a microprocessor executing a specific program, resident in firmware. The program allows a great flexibility, accepting, by initialization through an interactive sequence, the choice of the logical configuration selected by the operator. The interactivity is assured by a 8 colours CRT video display, an alphanumeric printer, and a keyboard. During the normal operation a graphic presentation of the status of the plant is displayed together with the alphanumeric messages selected by initialization.

At any moment, by operator's request, current values of variables or system messages may be displayed or printed. All data are periodically sampled, processed and dispatched to the appropriate preselected end-use: display, printer, RAMs, tape recorder, output gates to other systems. For the critical parameters two alarm thresholds are provided: overpassing the first one activates an acoustic alarm and generates diagnostic messages visualized on the CRT. Overpassing the second threshold generates new messages and, if necessary, issues blocking commands to turn off the faulty section(s).

The management sub-system includes a second microprocessor, with a ASCII keyboard, a printer, a BW CRT display. Operation mode is defined by the operator, using a high level programming language. All sub-routines for data acquisition and processing, and for commands, issuing, are stored in firmware: the program must only assure the "logical link" between these subroutines. The operator will be continuously informed on the operation performed by the processor, and specifically on the commands issued. In the first phase of plant operation the commands will be taken as suggestions by the operator, who will manually control the plant operation.

12. System integration and installation

Ground preparation is started on June 15,1983. A cable transportation system from the sea level to works area is being installed: the p.v. array will be completed by Dec.31, 1983 and operation of the system will start on Feb. 15, 1984. Acceptance tests may take place on March, 15, 1984.

Note: This report consists of the contributions from the following persons:

Mario Carlevaro (L'Aquila University)
Claudio Pelino (Technical Coordination)
Antonio Ponticello (L'Aquila University)

POWERING OF A SOLAR HEATED SWIMMING POOL

Location, country	: Chevetogne, BELGIUM
Rated peak power of array	: 63 kW
Contract number	: ESC-P-053-B (D)
Duration of contract	: March 1, 1981 - June 30, 1983
Total budget of project	: 1.218.500 ECU, CEC contribution 36,2 %, 441.000 ECU
Name of other funding bodies and thier share pf total cost	: Ministère des Affaires Wallonnes, : Prov. de Namur 31,2 %, 380.000 ECU SPPS - Premier Ministre 24,4 %, 297.500 ECU Contractor 8,2 %, 100.000 ECU
Contractor	: Association Momentannée IDE- ACEC (B) partners : Fabricable (B), ETCA (B)
Project Leader	: Marc VAN GYSEL IDE Parc Industriel B - 5430 Rochefort BELGIUM Telex : 41969 Phone (084) 21.37.71

Abstract :

The aim of this project is to implement a photovoltaïc plant that will supply electric power to satisfy two basic needs :

1° To drive the pumps of a 2.100 m2 solar thermal system that heats an outdoors swimming pool. This need is hourly in phase with PV and will be met with a stand-alone subplant.

2° to power the evening lighting of the pool.
This need is seasonnally in phase with PV and will be met by a battery aided subplant.

Both subplants can also deliver power to the local grid.
The system includes 3 modular 20 kVA inverter and a microprocessor-based control unit.

DOMAINE PROVINCIAL VALERY COUSIN – CHEVETOGNE
PROVINCE OF NAMUR
GENERAL VIEW OF THE SOLAR PLANT

1° INTRODUCTION

This 63 kWp photovoltaïc pilot plant is installed in the exquisite scenery of the Provincial Domain of Chevetogne.
70.000 People visited last year the Domain, which is situated in Province of Namur (Southern Belgium).

The Domain features 500 ha area , camping grounds, sports centres, and an olympic size outdoor swimming pool.

The swimming pool is heated by a 2100 m2 solar array, which is operated successfully since 1981.

The photovoltaïc plant is located near the swimming pool. It provides the power for actuating the pumps of the solar heating system and for lighting the vicinity at night.

2° GENERAL TECHNICAL DESCRIPTION OF THE PROJECT

The photovoltaïc plant is divided into two functional parts : a 40 kWp and a 23 kWp subplant. (see photograph of the single line diagram).

The two subplants have different purpose and operation, although their general design is similar.

Both subplants feature a photovoltaïc array connected to a 220 V nominal DC busbar.

On this busbar, batteries and inverters are connected. The inverters output to an AC busbar (220 - 130 V three phase) to which the AC loads and the grid can be connected.

The two systems can be paralleled at the DC and/or AC level.

The management system is common to both subplants.

The DC voltage level was chosen as a compromise between efficiency and availability of swithgear and other components.

The 40 kWp subplant is intended to provide electricity to the 14 circulator pumps of the solar heating system, totalizing 24 kW.

The actuation of these pumps is performed by the regulation system of the heating system so that the number of circulators switched on is roughly proportional to the irradiation.

The PV array is sized so that the output of the PV array fits bests with the energy demands of the circulators.

To compensate with non-linearities, a small buffer battery (120 Ah) is used in this system.

To avoid excessive charging rates, an additional degree of freedom for the regulation has been introduced by splitting the 40 kWp array in 3 sub-arrays that can be switched off by steps if the charging currents is too high.

Two inverters modules of 20 kVA each are connected to this subplant.

The 23 kWp subplant is intended to provide electricity to the lighting poles on the road leading to the pool and around it. This require a storage of 150 Ah for a day/night cycle, and one 20 kVA inverter module.

Both subplants can be connected to the local grid of the domain which own the plant and benefit the energy delivered. In this mode, which is foreseen to be used mainly in winter, the inverters can perform maximum power point tracking.

3° ARCHITECTURAL AND ENVIRONMENTAL ASPECTS

The area available for installing the array was a small hill east of the swimming pool.

The structures are designed so that module rows follow gently the curve of the hill, and, as can be seen on the photograph, the array does not disturb the skyline and is agreeably integrated in the landscape.

The natural grazing has been kept between the module rows.

4° OVERVIEW OF NEW TECHNOLOGICAL DEVELOPMENTS

- Modules

Special polarized plug-in connectors, were adapted to the standard IDE module, and allowed to reduce the installation labour.

- Module mismatch

Computerized handling of the electrical characteristics of the modules allowed to reduce the mismatch below 1 %.

- Structures

A two-pole structure was specially designed for the project. Its main feature is that the rows follow the slope of the terrain, reducing environmental impact and minimizing site preparation costs.

- Inverters

This project will be the first one to implement the high efficiency (90 - 95 %) modular inverters of 20 kVA manufactured by ETCA under an IDE development program.

- Control and monitoring

A microcomputer performs the monitoring and the automatic control of the whole plant.

5° MODULES

The modules are BELGOSOLAR 33-4 modules, manufactured by IDE. They feature :

- 33 W nominal power
- 36 monocristalline silicon cell
- Twin glass, EVA laminated encapsulation
- 2 shunt diodes.

They have been fully qualified by ISPRA Establishment (Spec 501) on June 15th, 1983.

As a special development for this project, they are fitted with 2 polarized plug-in connectors.

These connectors allowed the module wiring labour on site to be kept as low as half a minute per module.

The computerized filing of the electrical characteristics of the modules allowed to sort them according to maximum power point current.

The modules were arranged on the field so that the modules in one series string had nearly equal maximum power point current.

The procedure theoretically minimizes the mismatch, which is evaluated below 1 %.

6° ARRAY AND CABLING

The array consists of 124 series strings of 16 modules in series. (80 strings for hte 40 kWp subplants, 44 strings for the 23 kWp).

Each string is connected by an 5 mm2 underground cable to the control room.

The total wire and diode losses are below 2 %.

In the control room, a special panel is fitted with 124 3-positions-switches (one per string), which allow each string to be switched to conditions "operation", "short circuit", "open circuit".

In the "open" position, the string can be connected with a variable resistive load inserted in a special measurement gear.

This allows complete measurement of each string from the control room.

The measurement gear is portable, and can also be used to measure modules on the field.

7° ARRAY SUPPORT STRUCTURES

A special type of support structures has been designed for this projects.

As shown on the photograph, they are two-poles structures with a beam supporting the modules, whose frame provides part of the mechanical rigidity.

The beams are connected to the poles by articulations, so as to allow the rows to follow the slope of the terrain.

This has the effect of minimizing the environmental impact, and also requires a minimal site preparation.

The foundation works include :

- excavation with a drilling machine,
- shuttering of central holes,
- pouring of lean concrete,
- erection of the poles,
- grouting.

The erection of the poles and beams requires no special equipement. It was done by two men in a week.

The structures are made in hot dipped galvanized steel.
The beams are hollow so as to form a cable channel.
The structures were sized according the Belgian regulations.

8° LIGHTING PROTECTION AND GROUNDING

Statistical analysis led to the conclusion that no special protection was required against direct strokes.

A partial protection is provided by the 10 metre high wind measurements pole, which is properly earthed.

Against indirect strokes, a gas filled surge arrester is fitted on each polarity of the DC busbar right after the array breakers.

A buried grounding network is cabled on the field. It consists of 150 mm2 copper cable.

Each metal part of the array support structures is connected to this grounding network.

The module frames are grounded through their bolts.

9° POWER CONDITIONING AND BATTERIES

The power conditioning is performed by 3 DC/AC inverters.

Two inverters are connected in parallel in the 40 kWp subplant.

These inverters use MOSFET transistors. They are self-commutated.

The sine wave synthesis is performed by the PWM stored wave technique.

The units are controled by their own microprocessors, which perform :

- commutation
- synchronization with the grid
- master/slave control of paralleled units
- maximum power point tracking.

The efficiency is above 92 % at 10 % load, above 95 % at full load, including the filters, but excluding the transformer.

With the transformer, the efficency is more than 88 % at 10 % load.

The total harmonic distortion is below 3 %.

The batteries consists of two banks of 110 lead-acid cells, made by Oldham France (see photograph).

The 40 kWp subplant battery is 120 Ah, while the 23 kWp subplant is 1500 Ah.

10° MONITORING AND CONTROL SYSTEM

The monitoring system includes all sensors specified by the EEC experts, and an ECMA-43 recoder.

The monitoring system is controled by a HP 1000 computer and HP 2250 data acquisition system that also perform the automatic operation of the system.

Three levels of control are possible :

- manual
- manual with computer control
- full automatic

The operation modes of the two subplants are independant of each but some interlocking is possible.

Ten operating modes are implemented which are selected by push-buttons.

In automatic mode, each selection starts the proper switching sequence of actuators, and performs synchronization if necessary.

In manual mode, the operator pilots by reading the instruments. Only relay protection exists in this mode.

11° SYSTEM INTEGRATION AND INSTALLATION

All the subsystems were fully specified and ordered in 1982.

The foundation works for the array structure began in September and were completed in November 1982. This work started six month before the mounting of the modules so as to avoid frost while pouring concrete.

Meanwhile, the supporting strucures were manufactured by a local workshop (in Rochefort). The same people installed the structures on the site in December.

The manufacturing of the modules, of the inverters and of the main control cabinets started during the last quarter of 1983.

The underground cabling and the grounding network were also buried by the end of 1982.

The adaptation of the existing buildings to house the batteries and the control systems started end december and were finished in March 1983.

The main changes of the buildings were done to adapt to the safety regulations for the power cabinets and the batteries.

The batteries themselves were installed in March.

The existing switchyear was adapted also in March.

The module installation was foreseen to begin in March, but continuous rains delayed the erection of the fence, and precluded access to the field.

Modules installation started begin may, at which time the manufacturing of the modules was finished.

The main electrical cabinets were installed begin of May.

The computer arrived end of May.

The inverters were foreseen to be delivered begin of March.

The manufacturer had late development problems, and transported the inverters on site begin of May.

New problems in the operation of inverters were discovered and are not solved until today.

A recent schedule leads to November 1983.

For completionof the inverters and the final testing of the plant.

Meanwhile, all the other components including the monitoring system, and the accessible operating modes of the system have been tested satisfactorily.

The acceptance test took place between 22 and 24 June 1983.

12° CONCLUSION

The 63 kWp photovoltaïc pilot plant of Chevetogne has been installed end of 1982, and all equipment is installed and checked on June 30th 1983, save for the inverters.

The completion of the inverters and the commissioning of the plant is foreseen in November 1983.

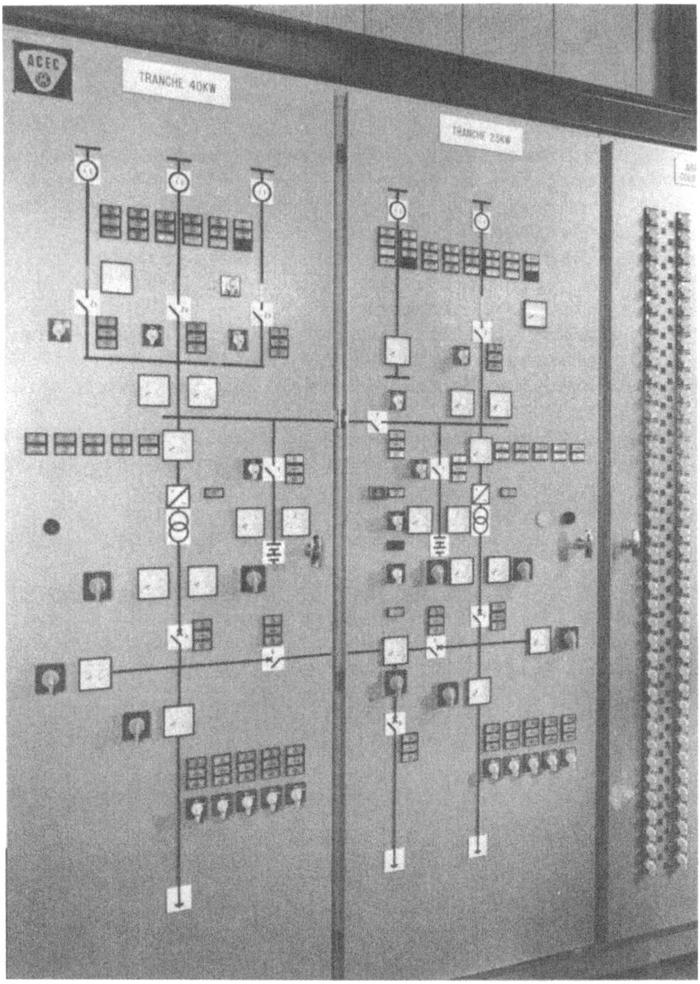

SINGLE LINE DIAGRAM
SYNOPTIC ON THE CONTROL CABINET

DOMAINE PROVINCIAL VALERY COUSIN – CHEVETOGNE
PROVINCE OF NAMUR
ARCHITECTURAL INTEGRATION OF THE SOLAR PLANT IN THE FIELD

SUPPORT STRUCTURES

STORAGE BATTERIES

RURAL ELECTRIFICATION IN FRENCH GUYANA

Location, country	: F.97390 Kaw, an isolated community in a French Overseas Dept. : La Guyane		
Rated peak power of array	: 35 kW		
Contract number	: ESC-P-060-F. (G)		
Duration of contract	: December 1981 - June 1983		
Total budget of project	: 816.700 ECU		
CEC Contribution	: 249.000 ECU + 49.000 ECU (bonus)		(30 %) (6 %)
Other funding bodies	:		
French AFME	294.000 ECU		(36 %)
Syndicat d'électrification rurale pour la Guyane	143.000 ECU		(18 %)
Members of the Consortium	: 81.700 ECU		(10 %)
Contractor	: Groupement Industriel pour le Générateur de Kaw SERI Renault Ingenierie (F) France Photon (F) Jeumont-Schneider (F) VARTA Batterie AG (FRG)		
Project Leader	: Bernard Aubert c/o SERI Renault Ingenierie 2 av. du Vieil Etang F 78390 Bois d'Arcy Telex : 695426 F Telephone : (33) 3 043 81 34		

Abstract

This project deals with a 35 kW photovoltaic generator for the rural electrification of the area of Kaw, an isolated community in the French Overseas Department "LA GUYANE". The photovoltaic generator has to replace two existing Diesel generating sets that supply Kaw at present. The system consists of : 492 new 72 watt modules from France Photon, grouped into 41 panels of 12 modules each. Supporting structures and panel wiring were optimized according to an overall approach of the system. A 410 kWh available capacity storage has been selected to ensure the continuous electricity supply throughout the year to homes in Kaw. Power is delivered at alternating 3 x 380 Volts + Neutral. The high efficiency inverter is a device specially developed by Jeumont Schneider using new transistors and the Pulse Width Modulation technology. Each subsystem is selected according to its reliability and autonomous operation which is necessary when supplying power to a remote zone. These subsystems (batteries with low self-discharge, power conditioning, inverter) as well as the high DC voltage allow this photovoltaic plant to obtain an excellent overall energy efficiency.

1. INTRODUCTION

This project deals with the electrification of a rural area remote from the electric distribution grid of the French overseas department "La Guyane".

The high installation cost per kilometer of a pole line distribution, and the rising operating and maintenance costs of Diesel generating sets about 2 ECU per kWh at Kaw, seems to make photovoltaic generators competitive.

The percentage of homes in the French overseas departments not connected to the grid is about 30 %. Such high percentages, are also valid in the major part of the mediterranean islands.

On the other hand, the improvement of living conditions is essential to halt the rural depopulation towards Cayenne, the main town, and to avoid the social and economical disadvantages of the rise of suburban overpopulation.

The wide potential for use of such photovoltaic generators, added to the demonstration of their effective operation as a reliable power source, are thus urgently needed.

We have, therefore, developed the qualities of reliability and autonomous operation for the 35 kW photovoltaic generator of Kaw.

2. GENERAL TECHNICAL DESCRIPTION OF THE PROJECT

At Kaw, the photovoltaic generator is composed of :

- the photovoltaic array of 41 panels of 12 France Photon's 72 Wp modules,
- the storage battery of 150 lead-acid Vartabloc 1500 Ah cells,
- the control-command device which comprises all the connections between the sub-systems, the necessary switches and safety devices, and the storage management system Varta "Logistronic",
- the all transistored 40 KVA inverter specially designed by Jeumont-Schneider (pulse width modulation technology),
- an automatic switch starts up a back-up diesel genset in case of failure of the solar generator,
- a building housing the control-command device, the inverter and the storage,
- the utilities (fence, ...).

The grid is supplied by the 3 x 380 V, 50 Hz produced current.

Each sub-system has been selected according to its reliability and automatic performance, which is necessary for operation in an isolated area.

The system structure is presented in the table below. Power management is essentially a control of the battery state-of-charge. The key role played by this parameter justifies a precise measurement of its value at any time.

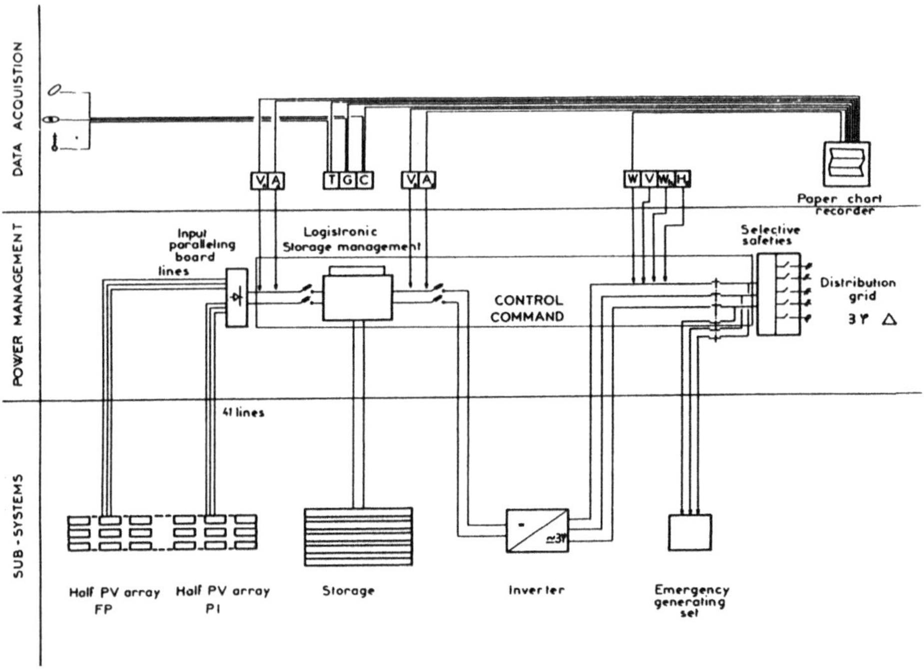

The general specifications of the main sub-systems are given in the table below :

P.V. ARRAY	Peak photovoltaic power 35 kW (installed, at 1 kW/m² insolation and 25°C Cells' temperature) Mean daily A.C. load supply 46 kWh in February 108 kWh in September <u>Composition</u> : - 41 panels of 12 - 72 W modules France Photon Tilt : 5°/horizontal
STORAGE BATTERY	Type : Lead-acid VARTA bloc Number of elements in series : 150 Capacity : 1500 Ah at C/10 hours Autonomy (at full load) - 2 days with an open-circuit array - 3 days under overcast sky conditions Voltage range : 292 V - 353 V New design : - maintenance free for 3 years - low selfdischarge : 3 %/month
INVERTER	Special "Solar" design for high efficiency and reliability Rated permanent power : 40 kVA Type : pulse width modulation transistored technology Output of inverter = 3 x 220 V at ± 10 % three phased A.C. Output of plant : 3 x 380 V with step-up transformer with neutral Frequency : 50 ± 1 Hz Efficiency at COS Ø = 1 of inverter/transformer 94 ± 2 % at full load 87 ∓ 2 % at a tenth of the full load Overload capacity : 100 % (permanent) No load consumption : - 200 W at the inverter output - 700 W at the step-up transformer output Conditions of use : - permanent operation (24 h/day) - temperature range 0 - 45°C - tropicalization : hygrometry 0 + 90 % marine atmosphere
OTHER FEATURES	- Fully automatic operation (no attendance needed) - Voltage regulation - Measurement of the battery state of charge by a new electronic device - System fully instrumented to allow all energy balances : Solar irradiation, DC energy generated, storage contribution, AC energy supplied...

3. LINK TO APPLICATION SYSTEM, ASSOCIATE GENERATOR

The supplied electricity is used by the inhabitants of Kaw for public lighting, lighting school, church, private homes, energy supply to refrigerators and cold stores (for meat and fish only).

The Photovoltaic generator is located at about 100 meters from the village. An extension of the existing pole line has been made by the national French utility "Electricité de France", from the diesel engine generator to the Photovoltaic plant. The two existing diesel gensets have been replaced by a new one as a back-up by EDF. Then, some precautions have been taken :

- it has been made physically impossible to connect in parallel the Diesel generated and the photovoltaic generated currents,
- a repartition, as equal as possible, of the appliances and users among the 3 x 380 V phases,
- a "selectivity of the safety devices" against the short-circuits in order to prevent the shut-down of the inverter, rather than the melting of the user's fuses.

Morever, some advice has been given to the inhabitants, to save the electricity to allow the solar plant to supply the grid at 100 % all year round. In particular, the incandescent bulbs of the public lighting have been exchanged for low consumption fluorescent lamps.

4. ARCHITECTURAL AND ENVIRONMENTAL ASPECTS

Kaw is a small village situated in a small thicket in the middle of a Savanna zone, a huge clearing hidden in the equatorial forest of the French overseas department of "La Guyane".

Kaw can be reached by 80 km of track leading on from the few km of road from Cayenne, the main town. The last few kilometres must be covered by canoe, as the track stops at the edge of the forest.

The 70 inhabitants live by agriculture (hearts of palm, tropical fruits) and by a small amount of stock rearing, hunting and fishing.

We cleared a zone of forest bordering the village to erect the photovoltaic generator, while trying to preserve possible zones for Kaw's extension, as well as future access road amenities.

The building housing the storage battery and control-command device has been built in the same style as the other village buildings. The four-sided roof amply overlaps the pebble-dashed walls, painted in bright ocre. The canopy thus protects the building from direct sunlight and preserves a dry area around the building when the heavy rains, common in this region, fall. The interior layout and a grillwork of large holes has been provided enabling the Eastern Sector Trade winds to penetrate.

Solar Source

Like tropical zones elsewhere, the climate of Kaw can be divided into a wet period from February to May, a dry season (June, January). However, as it can be seen in the figures below, the discrepancy between wet and dry season is not so strongly marked from the point of view of the solar overall irradiation. As a consequence, the full energy supply to the village can be obtained by a solar generator throughout the year, without a dramatic overdimensioning for the dry season to be sure of supplying enough power during the wet season.

G	J	F	M	A	M	J	J	A	S	O	N	D	Avg
$kWh.m^{-2}$	3.48	2.95	3.19	3.13	3.08	3.80	4.74	5.07	5.81	5.61	4.54	3.32	4.06

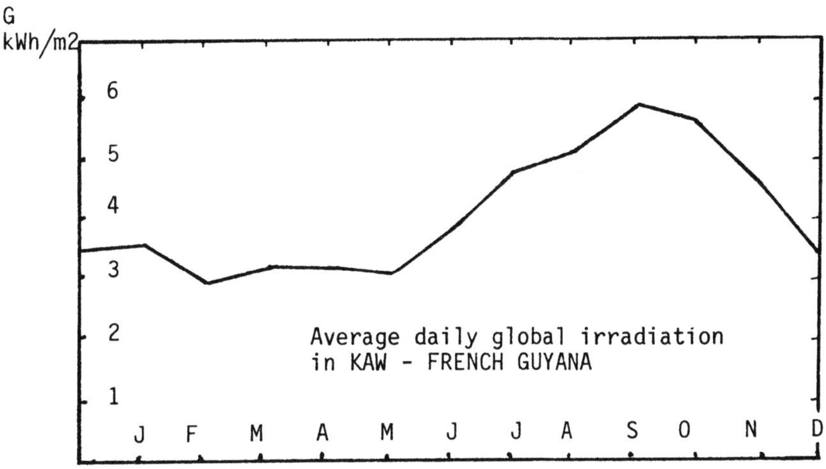

5. A REVIEW OF NEW TECHNOLOGICAL DEVELOPMENTS

In our efforts at technological development, our main aim has been to improve the quality per price of future photovoltaic generators. Thus, we have particularly studied the following :

- the supporting structures, in order to obtain the lowest possible cost for the operating unit, including manufacture, transport and site assembly,

- the inverter, in such a way as to obtain the best possible yields whatever the load, which is very variable in such cases. The inverter's yield, the last link in the photovoltaic chain, is, in fact, of prime importance in establishing the most accurate dimensions for the other sub-systems,

- the knowledge of the battery state of charge, in order to improve its length of life. We have thus developed an integration system for the electricity passing through the battery.

6. MODULES

The modules used are the FP72, made by the French company FRANCE-PHOTON. They are composed of 68 cells of monocristalline silicon of diameter 100 mm. The encapsulation involves a front side in glass and a rear side in silicon resin.

The electric outlet is composed of waterproof sockets of Amphenol type.

The modules include the necessary safety parallel diodes.

7. ARRAY AND CABLING

The output voltage of the photovoltaic array is 330 V, the modules are thus, grouped into series of 12 making up a "panel" and the "panels" making up the array are parallel grouped.

The 12 panel modules are connected by a "garland" on to which each module is connected by its Amphenol socket. The "garland" leads to a connecting box fixed on to one of the panels' supporting feet. Each panel is individually linked to the "paralleling board" in the control cabinet. The panel safety diode-series are also situated here. Thus, the tension of each panel can be individually checked from the control cabinet, which therefore allows any possible cabling or module failures to be detected.

In the same way, the "intra-panel" cabling system enables each module to be easily disconnected and checked.

The installation of the "garland" and cabling of the 12 panel modules can also be carried out in less than 10 minutes by 2 unskilled operators.

The matching of modules, the choice of section area of cables and the diodes lead us to believe that all the cabling losses and dispersion of module characteristics are below 2 %.

8. ARRAY SUPPORT STRUCTURE

The structures supporting the modules have been perfected with the aim of reducing, as much as possible, the costs of all manufacturing - assembly - installation, including the civil works necessary for the foundations. Moreover, it was necessary to dispense with lifting gears due to access problems to Kaw. All structural parts have been transported by hand from the landing-stage on the bank of the canal.

Therefore, the support equipment of the 12 modules (approximately 860 Wc and 10 m²) only includes :

- 2 footings each assembled on a concrete block by two threaded rods,
- 1 central double beam and 2 bracketing sections which fix the modules on to the footings,
- 2 threaded rods join the modules and the metallic beams.

Individually, the structure elements weigh less than 25 kg.

To enable the rapid growing tropical vegetation to be regularly cut, the panel is situated 1,5 m from the ground in order to facilitate access to the area below the modules.

The panel slopes 5° from the horizontal plane towards the South, in order to facilitate evacuation of rain water.

A duct between the two beams of the "central double beam" is provided for the" garland" connecting the modules to the connecting box.

9. LIGHTNING PROTECTION

French Guyana rarely experiences lightning.

The maximum keraunic level (TD) is 30 days per year (source : European Space Center). Since Kaw is near the coast, it can be expected that this average value exceeds the "spot" value for the site. Calculations, carried out according to a method recommended by the CEC, have shown that, for the Kaw photovoltaic generator, the average time between two lightning strikes hitting the installation is in the order of 300 years. As the estimated life span of the solar plant is 20 years, protection against lightning has not been judged to be of critical importance.

The only measure taken is a careful earthing for the 41 structures through a 2 x 30 mm² cross section copper wire that links the whole array.

10. POWER CONDITIONING AND BATTERIES

The power conditioning system is composed essentially of two parts : the inverter, supplying alternating current in 3 x 380 V, and the battery charge monitoring system.

The Inverter :

This was specially designed by Jeumont-Schneider for use in solar installations. The highest possible efficiency has been sought, whatever the load. The inverter is linked to an outlet transformer area so as to minimise losses. The inverter, fully transistorized and using PWM technology, because of the way it has been built, without a filter it has no harmonic until the 11th row. However, to keep the harmonic distortion ratio below 6 %, it has been necessary to add a filter, which has thus lead to a slight lowering of overall efficiency and a small increase in no-load losses.

The output voltage, the image of the input voltage of the inverter is maintained at 380 V ± 10 % when the inlet voltage varies from 292 V to 363 V. It is thus necessary to use a battery with internal weak resistance so as to minimize the variation of its voltage in between charge and discharge.

The efficiency curve of the Kaw inverter, measured at the transformer terminals, is given below as a function of the load.

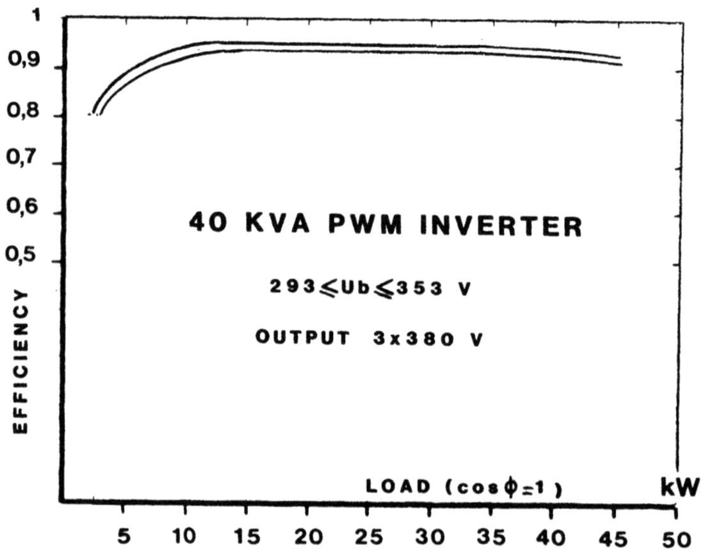

The battery state-of-charge monitoring system :

To prolong the life of the battery, it is necessary to limit the depth of discharge to 70 % of the maximum load and to avoid overloads.

With these purposes in mind, VARTA has developed the LOGISTRONIC. This system, taking into account the various parameters (type and temperature of the battery, voltage), enables the quantities of electricity entering and leaving to be measured by a microprocessor and thus giving the state-of-charge. The device emits electric signals when the high and low charge limits are reached, thus ordering the photovoltaic system to be put into the "battery overload" position, or to stop the inverter and current supply to Kaw's inhabitants, respectively.

The battery :

The battery is composed of 150 of the new VARTABLOC cells and has been specially designed to permit the slow discharge regimes, characteristic of photovoltaic generators for the electrification of isolated villages.

All measures have been taken to keep maintenance operations to the minimum and to prolong the life of the battery :

- large electrolyte reserves,

- careful sealing to avoid acid leaks,

- catalytic plugs recombining the oxygen and hydrogen possibly disengaged.

11. MONITORING SYSTEM

The Data monitoring system now installed gives, on a paper chart recorder, the variation of the meteorological and electrical parameters of the installation. These variations must be known, in order to fully understand the system's functioning but also in the case of a breakdown to discover the faulty organ.

The sum of the daily total solar energy is made and the electric energy supplied to the village recorded.

This system enables calculations to be made of the efficiency of all the photovoltaic generator sub-systems, array, battery, inverter.

Moreover, to facilitate the processing of data by the ISPRA Joint Research Center, we are at the moment preparing to add a magnetic recorder and some extra gauges :

- strain gauges on the support structures,

- anemometer at 10 m.

12. SYSTEM INTEGRATION AND INSTALLATION

After the clearing of forest by the Kaw inhabitants, we carried out the necessary civil works limiting ourselves to the construction of the supporting blocks, the panels and building housing the battery, the power conditioning and the monitoring system during the months of October and November 1982.

At the same time, we installed the photovoltaic array field and the cabling. The system was first put into operation at the end of November and on the 25th November the village received photovoltaic electricity (for the first time).

Trial runs and inspections of the system followed on from this.

During this time, modifications to the existing network were made, in particular the introduction of safety devices necessary to maintain the smooth running of the solar plant.

On the 11th January 1983, Electricité de France approved the photovoltaic generator, linking it definitively to the network.

On this occasion, various satisfactory tests on the resistance to short circuits in the subscribers home, on the line and on the plant terminals were made.

13. CONCLUSIONS AND FIRST RESULTS

From the 11th January 1983 until the end of June, 15 MWh were supplied to Kaw in 4000 operating hours. The wet season (from November to the end of July) passed, causing only a few stoppages (6) due to lack of sunlight.

In June, the average consumption was 105 kWh under good meteorological conditions (the dry season is early this year in comparison to 1982). See next page - the typical load profile recorded at KAW on June 21.,1983.

The only problems occuring have been due to the breakdown of two microswitches which have now been replaced.

We are therefore hopeful that the Kaw photovoltaic generator will show that solar plants offer an efficient and reliable source of electricity for isolated sites.

For obvious economic and socio-political reasons, the need to improve the living conditions of inhabitants of dispersed rural villages all over the world is very great : photovoltaic electricity, we believe, represents a major contribution towards fulfilling this objective.

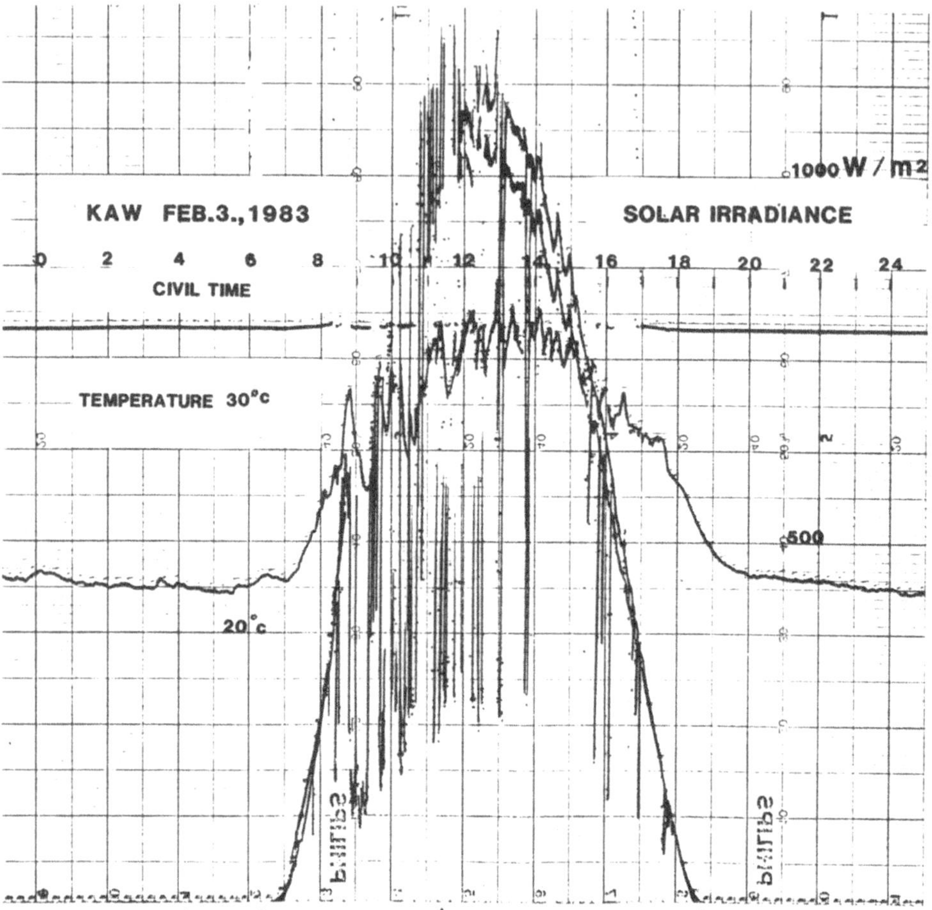

The input : the solar energy ⟶ ⟵The output : the village load

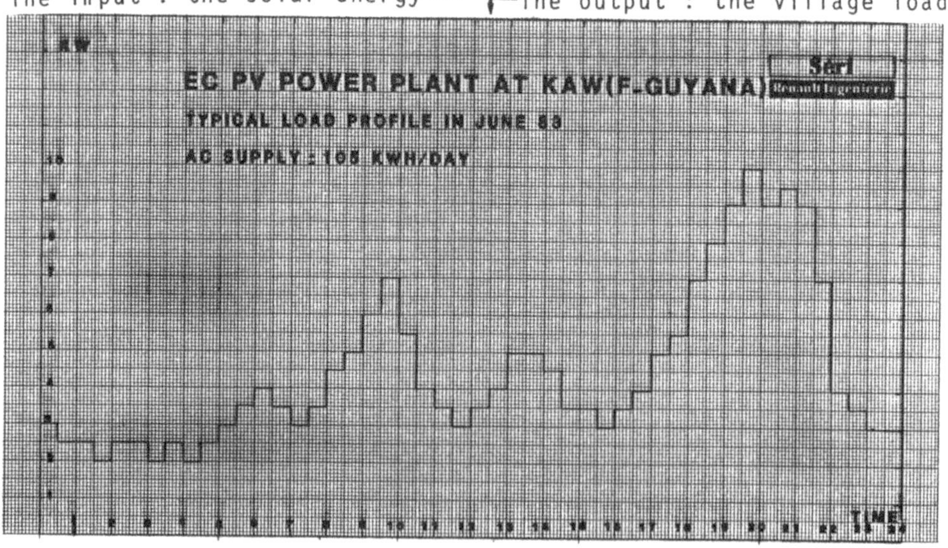

POWER SUPPLY FOR TV AND FM EMITTERS

Location, country	MONT BOUQUET, FRANCE
Rated peak power of the array	50 KW
Contract number	E.S.C. - P - 059 - F (D)
Duration of contract	MARCH 1, 1981 - JUNE 30, 1983
Total budget of project	1 050 000 ECU CEC CONTRIBUTION 360 000 ECU (35 %)
Name of other funding bodies and their share pf total cost	COMES - 480 000 ECU (45,7 %) PHOTOWATT - 105 000 ECU (10 %) TDF - 105 000 ECU (10 %)
Contractor	PHOTOWATT INTERNATIONAL S.A. 131, route de l'Empereur 92500 RUEIL MALMAISON Tel : (1) 708 05 05
Project leader	Patrice COUREAU PHOTOWATT INTERNATIONAL S.A.

Abstract

MONT BOUQUET project experiments the use of a 50 kWp photovoltaic system to power FM and TV emitters of TELEDIFFUSION DE FRANCE which are located in the south of France on a remote mountain near Alès. The emitters are using AC triphased current and will be powered either by the photovoltaic plant while solar power is available, or by the grid during the night and bad weather periods. The load constitued by the emitters will be divided into six separate su-loads which will be switched on or off on account of the shape of the insolation of the day by the mean of a load adaptator.

A battery of low capacity will fix the limits of the input voltage of the 30 kVA PWM inverter while storing excedent of energy coming from the array and not immediatly used.

A specific charge level controller will regulate the charging mode of the battery by the mean of a "multi cut-off" system allowing a thin adjustment of the charge.

A separate monitoring system will collect meteorological and energetic data so that to notice the operating of the station and calculate the global efficiency.

1. INTRODUCTION

The experimental photovoltaic power station, of the emitting center of "Telediffusion de France" of Ales-Mont Bouquet, is one of the fifteen pilot units of the photovoltaic scheme, launched in 1980 by the European Communities.

This is one of the 4 installations settled on the French territory within this scheme, with :

- the international Nice-Côte d'Azur airport
- the village of PAOMIA, Corsica
- the village of KAW, French Guyana.

PHOTOWATT International S.A. supervised the complete implementation of this station, as well as that of the station of Nice-Côte d'Azur.

2. GENERALS

The scope of the project is to use solar photovoltaic energy to partially power an emitter of Telediffusion de France, situated about 20 km of Ales (Gard, South East France), on a rocky summit named Mont Bouquet, culminating at 607 mètres.

The center of Ales-Mont Bouquet captates the emissions from the Montpellier-St Baudille center, and pilots several local re-emitters in order to provide emissions for the numerous shaded areas due to the local very hilly geography.

The area used to settle the photovoltaic captors field is situated on the edge of a cliff, and forms a band of approximately 15 metres large and 140 metres long, oriented right south. The 6-degree southward tilt angle allowed inter-row distance to be reduced to 3,70 metres.

The so formed photovoltaic field is divided into 70 branches of ten 72 peak-Watt modules, cabled in series. The last branch, cabled in 48 volts, serves to power all the measuring, watching, and data acquisitions sections of the station.

Each of the 70 branches is connected to a battery charge regulator, monitoring the disconnection of some or all branches according to batteries charge level. Regulation of the device is operating accordding to the voltage and temperature of the batteries elements.

The emitting center consists of 3 emitters - radio and television - and a number of other commodities. Total power consumed by the center fluctuates between 35 and 40 kVA, with an annual consumption of 256.000 kWH.

Allowing a complete autonomy of the center with only photovoltaic energy would have meant a very large power station that, although technically very possible, would have been quite different from future generators which will have to match the low-consumption emitters presently under development.

The Ales-Mont Bouquet center has indeed been installed more than 10 years ago, and the equipments were designed without taking in consideration the reduction of energy consumption, as a line of the national grid was available on the site. For these reasons, it has been decided to install a station of only 50 peak kilo watts, representative of the future power station, to test the behaviour and functioning of such unit.

The photovoltaic field of 50 kWp will thus supply, according to local meteorological conditions prevailing in the nearby city of Carpentras, 48 000 Kilowatt-hour per year, giving a net available energy of 37830 kWh/year, considering the yields of the ancillary equipments namely the charge regulator, the batteries and the converter. Net share of photovoltaic energy will thus average 15 % of total energy consumed.

3. STATION CONFIGURATION

(chart n° 1)

The part "captation of solar energy" is made out of 710 photovoltaic modules of 72 peak Watts (mean value), and shows an active area (silicon cells) of 700 m2, settled on an area of 2000 m2.

Total peak power of the field reaches 50 kiloWatts (measured according to international standard).

Battery bank is made out of 106 elements in series, of 800 Amps-hours, assuring a nominal voltage of 212 volts.

The protection cabinet includes all the circuit cutting elements, able to isolate each of the sub-systems in case of failure or incident, and allow intervention with all security.

Conversion of electrical direct current into alternating current is made thanks to an electronical converter (chopping type), delivering a 3-phase 380 volt current, able to power the 5 charges which constitute the emitting center.

Considering the comparatively small size of the accumulator bank, and in order not to waste the energy produced by the photovoltaic field, a system of progressive and sequential putting into service of the 5 charges has been designed. This allows the consumption of some or all of the 5 charges to match the energy produced by the photovoltaic field, according to the insolation level.

A number of steps have been taken in order to guarantee the lifetime of the electronic equipment and the modules, in case of lightning strikes nearby the photovoltaic field.

The whole unit is built according to the prevailing standards, and includes all the requested protection as, especially, an isolation control unit to watch the whole installation.

4. WORKING PRINCIPLES

The emitting center consists of 5 separable charges as follows :

		P	E
- TV2 (Channel 2)	working from 9:30 A.M. to midnight	8 kVA	116 kWh
- TV4 (Channel 1)	working from 9:30 A.M. to midnight	9 kVA	130 kWh
	low service : midnight to 9:30 A.M.	7 kVA	6 kWh
- TV3 (Channel 3)	working from 1:45 P.M. to midnight	8 kVA	80 kWh
	low service : midnight to 1:45 P.M.	6 kVA	84 kWh
- FM (3 emitters)	working from 6:00 A.M. to 1:00 A.M.	3 kVA mean value	57 kWh
- Commodities	24h/day, from 4 to 10 kVA	7 kVA	168 kWh
			702 kWh per day

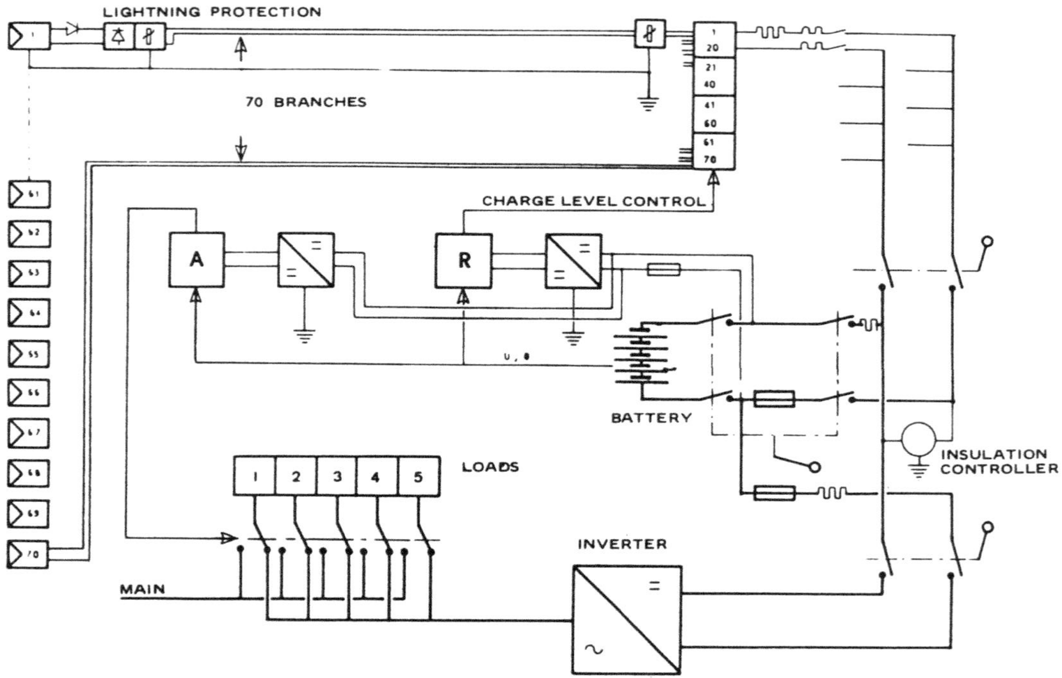

MONT BOUQUET PROJECT
PHOTOWATT INTERNATIONAL S.A.

<u>chart 1</u>

Profile of consumption may thus be represented as follows :

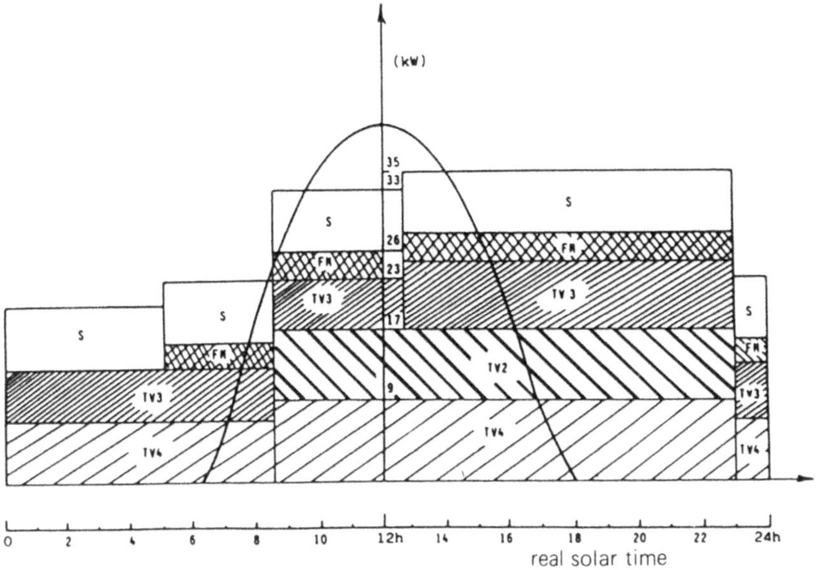

Maximum available power from the photovoltaic field corresponds to some 40 kW, which is very near the maximum needed power for the whole center.
Practically, this value is only reached at noon (solar time) with very favroable conditions.

The above diagram obviously shows the necessity of connection to the main, to guarantee full-time function of the emitting center.

Two types of function were foreseen :

- First, it would have been possible to power permanently only 1 of the 5 charges, i.e. 5 or 6 kW during the whole year. in this case, the batteries would have to be sized in order to stock excess energy of favorable periods, to compensate for winter deficit periods.
  This would have led to a quite gigantic bank of batteries, pretty apart from the standard power station of the future.
- Considerations of available space, cost and maintenance have oriented the choice towards a second solution, very similar to a use of photovoltaic energy known as "following the sun" thus reducing the bufferstock to a minimum.

Its principle is the following :

- Connection of each utility to photovoltaic is switched on as the amount of energy from the solar modules becomes sufficient. The surplus energy is then fed to the buffer-batteries, while waiting for the next switching on.

The excess energy produced in mid-day is also stocked in the available remaining capacity of the batteries, and restituted in the afternoon before each switching off. Same process occurs each night in order to reiterate an identical cycle on the next day.

The cycling of the battery, in normal function, is settled between 20 and 25 % of the nominal capacity (as measured at 10 hours discharge rate).

It should be noted that the diagram assumes a working day with ideal sun, and though the converter might seem to be a power limit to the system, consideration of the normal - less optimistic - meteorological conditions, has led to choose this peak power - 30 kVA - as the capacity of the inverter.

In case the main is faulty, or in case of lengthy cuts of the main, the photvoltaic system assures automatically the power to the entire center, but with the emitters working half-power, during the time of the cut. Normal function starts again as soon as the main becomes available.

The diesel|alternator group, previously installed on site, is put into service only as the last possibility, in case the batteries reach a critically low charge level.

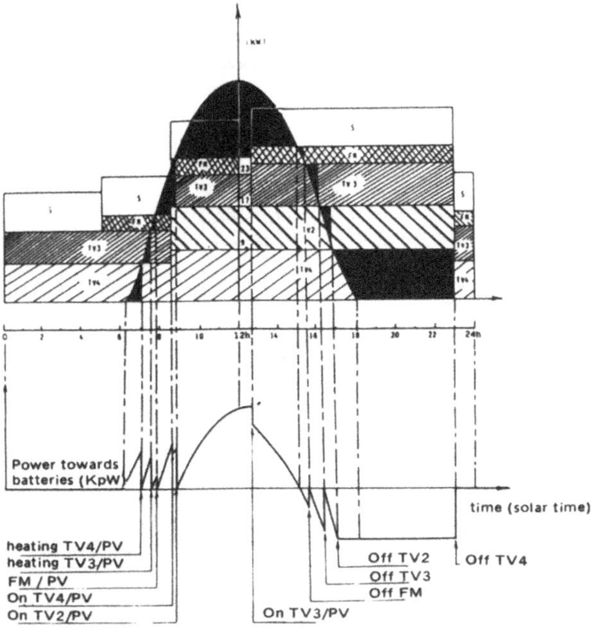

5. DESCRIPTION OF SUB-SYSTEMS

5.1. Modules

- Made by PHOTOWATT INTERNATIONAL S.A.
- Type PWP 800
- Peak power 73 W
- Voltage 12 or 24 V
- laminated technology (glass/PVB/tedlar^R/alu/tedlar^R)

- mono cristallin silicon cells, Ø 100 mm
- metallic connection box sticked at rear of module
- wire connections with "fast on" plugs
- protection against "hot spot" phenomena by 6 diodes settled in connecting box
- self-bearing aluminum frame
- installation with 4 bolts, from front face of module

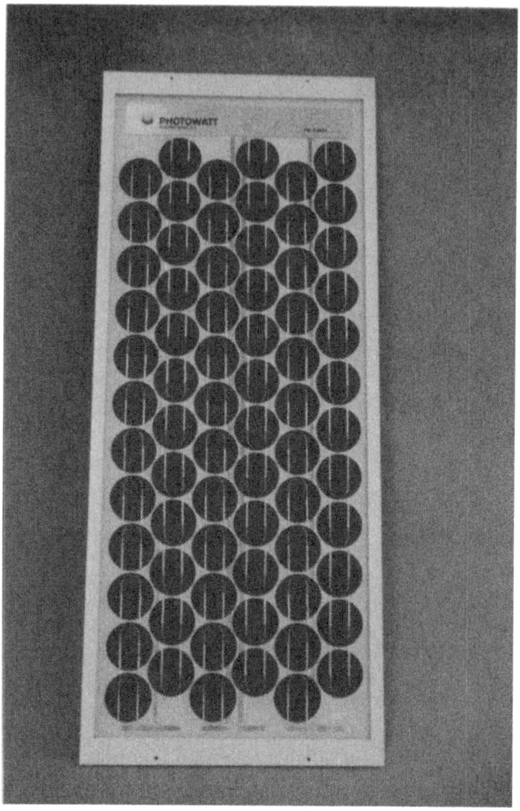

5.2. Metallic structures

Each branch of the photovoltaic field is made of 2 panels of 5 modules.

While developing light and adaptable structures suitable for different configurations of panels and various tilt angles, the following considerations have been taken into account :

- no redundant parts between frames and modules
- no necessity to bore holes on site
- no electrochemical pile phenomenon
- little manpower needs for erection

The design of the metallic frame thus uses the frame of the module as part of the structure itself.
The unit is principally made of 2 high and low girders in extruded aluminum. Their profile has been especially studied to allow setting of a rear "pi"-like profile, studied to abide forces created by winds up to 180 km/h. The whole structure is mounted upon 4 tubes in galvanized steel, anchored into the ground through shell-like aluminum articulation pieces, allowing perfect adjustment of height and tilt angle without need to make precise foundations.

these structures were designed and tested
according to the French standards for
buildings construction

5.3. Cabling

The nominal working voltage is set at 220 volts, continuous, to take into account unavoidable charge losses in cables, due to the distance between the photovoltaic field and the technical building, as well as the length of the photovoltaic field itself. The length of cables varies from 80 to 250 metres.

The relatively high tension means that all metallic parts have to be linked to the ground. In this respect, a copper strip, burried and linked to the ground grid of the emitting center, is connected to the branches by alumino-thermical osoldering. This strip also serves to evacuate currents induced into the cables by lightnings.

The 71 cables, connecting the branches of the photovoltaic field to the battery charge regulator, are settled in a prefabricated concrete through set onto the ground an aligned at the extremity of the 36 rows of panels.

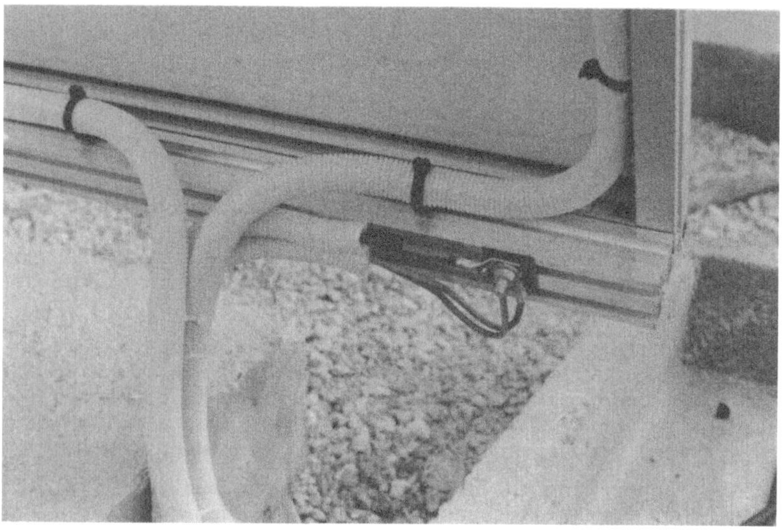

5.4. Lightning protection

Mont Bouquet is a high point within a relatively tormented relief. Moreover, the antenna upon the emitting post of Telediffusion de France particularly exposes the site to lightnings in case of storms.

Protecting against direct strike of lightning is practically impossible with such a surface, and the essential step taken was to make very carefully all connections of the different metallic parts to the ground.

On the other hand, currents indirectly induced into the cables following lightnings are an important risk.

For this reason, a specific modular protection, adapted to the problem, has been developed by "Soule EM".

The basic principle of these units is not to set any polarity compared to the ground, but to limit over-voltages between ground and phases, with 3 adapted varistances, situated at both extremities of each cable.

An additional protection, on the module side, allows to protect the diodes mentioned here above.

Choosing and sorting the varistance is of utmost importance, considering high number of modules and the open-circuit voltage, which overpasses 400 Volts.

5.5. Charge regulator with sequential monitoring

The essential aim of a charge regulator is to allow the stocking of energy in batteries, in such a way as to preserve their normal lifetime.

Thus it has to limit the in-coming current in the battery, while maintaining the voltage below a pre-determined threshold, which itself evoluates according to temperature. Any overpassing of this voltage would cause damage to the plates, and an over-consumption of water.

A number of solutions are available to fulfill this function, and PHOTOWATT INTERNATIONAL S.A. has specifically developped an original cutting system, allowing to palliate the drawbacks of classical systems.

The principle relies on the comparison, every 30 seconds, of the working point of the batteries with a pre-set temperature compensated data. Result of the comparison induces the putting on or off of one of the branches of the photovoltaic field.
The maximum current increase or decrease is thus 2,5 Amp and hence it is not necessary to provide for the hysterisis phenomenon.

Hence, batteries charging is achieved in optimal conditions. The commuting parts are made of special relays, placed in series on each branch of the photovoltaic field. The sequence monitoring is made through logic cards assuring an overall consumption of about 30 Watts.

5.6. Accumulator bank

In order to guarantee the working life-time of 8 to 10 years, the cycling conditions prescribed to the batteries should not exceed 25 % of the capacity. Choice has been set on OLDHAM's lead elements, showing little content of antimony, and having the following characteristics :

- 106 elements, SYT8 type, in series
- capacity : 800 Amp hours at 10 hours discharge rate
- low antimony content
- tubular positive plates, with oxyde added
- high electrolyte level
- nominal voltage : 212 Volts
- low sel-discharge rate
- weight : around 7 tons

5.7. Conversion continuous / alternating

The continuous/alternating conversion is achieved thanks to a high performance converter, derived from the standard range developed by "Aerospaciale". it has indeed been necessary to take into account the continuous in-voltage delivered by the installation. The main features are :

- power : 30 kVA
- autonomous inverter, 3-phase, 380 Volts with neutral out
- chopping with impulse with modulation transistors
- commutation of filters at 3 kVA to improve yield at low charge
- entrance voltage : from 180 to 260 V, continuous

- out-voltage tolerances . 380 V ± 1 %
  . distorsion less than 4 % with non deforming charge
  . frequency 50 Hz ± 1 %

within the whole range of entrance voltage and/or charge power

- current limitation : 1,25 Inom. during time from 1 to 100 sec.
- yield of the inverter, with cos φ = 1 is shown on following curve.

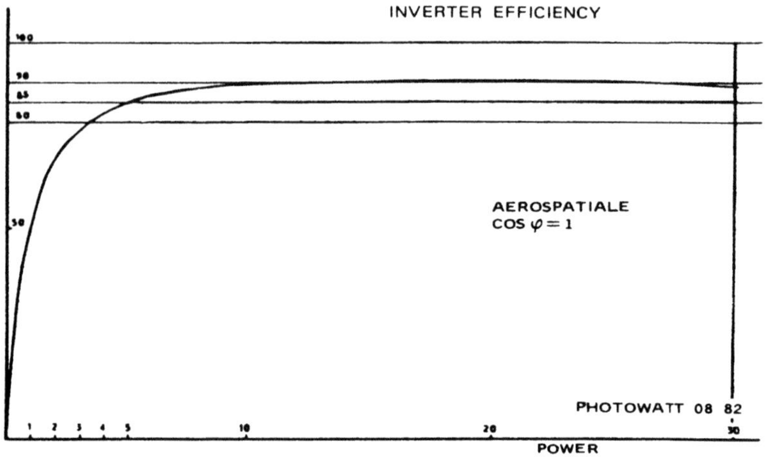

5.8. Linkage with low voltage board

Telediffusion de France has provided for the replacement of the former low voltage board, the modifications of which would have been pretty difficult. Taking into account, in the studying phase, all the characteristics of photovoltaic energy and of charge regulation equipments led to a practically immediate putting into service of this device.

6. CIVIL WORKS

Mont Bouquet is a locally appreciated touristic site, and the panels area was previously a parking lot.

The installation of the power station has thus deprived the site of that natural parking and of part of the access road which had to be rebuilt.

The preparation of the site consisted in an East-West levelling, following the natural tilt of the field.

Anchoring to the ground the 568 settling tubes was made with concrete blocks, casted directly onto the ground, a mobile jig providing a correct setting of the tubes.

This solution was chosen after a technico-economical comparison with another solution : boring 568 holes of around 40 cm diameter. Nowever, the nature of the ground (rocks and man-added materials) would have made that job an expensive one. On this specific site, casting concrete blocks was the most economical solution.

A small building has been especially erected at the southern extremity to house all the batteries.

A collection of cables is made with a concrete trough as previously mentionned.

7. ASSEMBLING - TRANSPORTATION - ERECTION

The desing of supporting metallic structures allowed the transportations of pre-assembled panels. This possibility was taken advantage of, since the 142 panels were assembled in the factory, pre-cabled, and checked with specific mounting kits.
This operation led to a substantial reduction of mounting time on site. Panels were transported to site by lorry, and settled within 9 days by M.G. FRANCE.

8. DATA ACQUISITION AND PROCESSING

One of the objectives of European launchers of these photovoltaic power station projects is to evaluate the evolution of the performances of each project with time.
Hence it was necessary to equip each power station with relatively complex measuring equipments, and to deliver informations suitable for comparison.

The basic principle leading to the definition of the sytem is to consider this measuring unit as completely independent from the power sation, in other words excluding any active role in its functioning.

The principal aim of this unit is to measure all data regarding energy, meteorology and state of generators, to put those in exploitable shape, then to visualize them on a synoptic panel while being tape-recorded. This resulted in the configuration shown on following page.

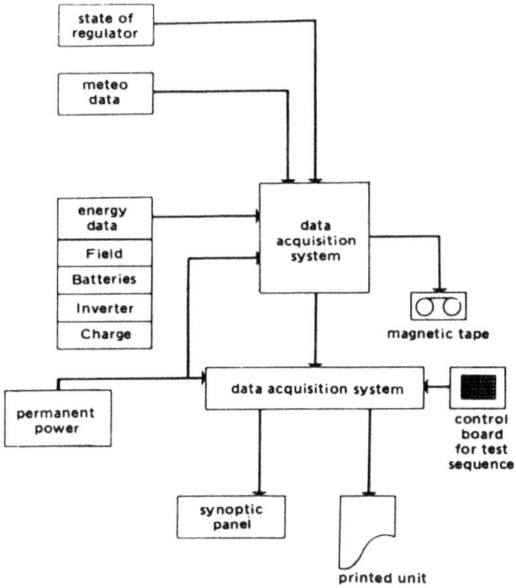

The 71st branch of the photovoltaic field, cabled in 48 Volts, with its independent battery, is an autonomous generator assuring power, without external help, for the entire monitoring system which is thus completely separated from the main generator.

Data are measured with isolated captors, and forwarded to a first sub-unit which transforms the informations into a digital form, and makes the mathematical calculations to supply mean values, extremes or integrals upon give time periods, and which registers these data on a DC 300 XL cassette in a pre-determined form, common to the 15 projects.

These intermediate informations are taken from this first module towards the next one which pilots a visualization board, for instant or cumulated informations, and a printing unit. It also proceeds to testing each photovoltaic branch.

MEASURED PARAMETERS

meterological	. global horizontal illumination
	. diffused illumination
	. illumination on the panels plan
	. outside temperature
	. wind speed
	. wind direction
	. effort on structure
state of generator	. number of branches in service

ENERGY DATA

field	. tension of 1 branch
	. tension of 1 Module
	. current delivered
battery	. tension
	. current
inverter	. current in
	. current out
charges	. charges connection
	. power taken from main

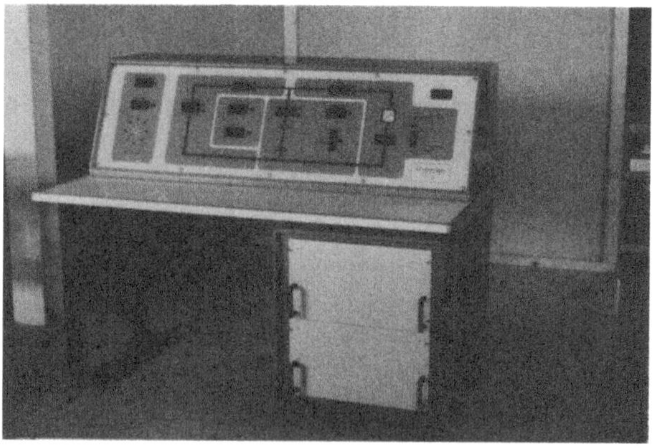

Measuring these informations is done every second, and registration every hour

9. WORKS SCHEDULE

The main phases of the installation of Mont Bouquet power station were as follows :

- study phase started January 1981
- design and development of components : course of 1981
- manufacture of equipments : January to June 1982
- pre-assembling of panels in factory : July and August 1982
- civil works : March to June 1982
- first cabling works : July/August 1982
- mounting, cabling of panels : September 1982
- start of power conversion parts : October 1982
- setting and start of charge regulation systems until December 1982
- setting and start of monitoring systems : March 1983
- reception : early April 1983

10. ECONOMICAL ASPECTS

The installation of a power staion of the size of Mont Bouquet allows to evaluate the respective weight, cost wise, of the different parts of the unit, as well as to split the respective parts of equipment and of research and development works.

This is shown in the following chart :

	TOTAL	material	R ¤ D
Modules	41 %	100 %	0
Frames / cabling	3 %	80 %	20 %
Energy transfer	3 %	90 %	10 %
Regulation / protection	6 %	70 %	30 %
Batteries	4 %	95 %	5 %
Conversion DC to AC	8 %	90 %	10 %
Civil works	13 %	95 %	5 %
Transportation / erection	10 %	80 %	20 %
Monitoring	5 %	60 %	40 %
Miscellaneous	7 %	60 %	40 %
	100 %		

Hence it appears that, apart from the modules part which remains an important one, major parts pertain to civil works and transportation/ erection, representing nearly a quarter of the total.
This point particularly stresses the importance of the choice of structures, their means of anchoring to the ground, and the possibility to pre-assemble the panels in factory allowing a limitation of interventions on site.

NICE AIRPORT SURVEY AND CONTROL

Location, country	NICE, FRANCE
Rated peak power of array	50 KW
Duration of contract	DECEMBER 11, 1981 - JUNE 30, 1983
Total budget of project	1 020 000 ECU CEC CONTRIBUTION 360 000 ECU (35 %)
Name of other funding bodies and their share pf otal cost	COMES - 493 000 ECU (48 %) CCI - 67 000 ECU (7 %) PHOTOWATT - 100 000 ECU (10 %)
Contractor	PHOTOWATT INTERNATIONAL S.A. 131, route de l'Empereur 92500 RUEIL MALMAISON Tel : (1) 708 05 05
Project leader	Patrice COUREAU PHOTOWATT INTERNTIONAL S.A.

Abstract

This project proposes to use a photovoltaic power station installed on the roof of the freight building of NICE COTE D'AZUR AIRPORT to supply AC current to main apparatus of the NAVAID system.
The aim of this equipment is to monitore all the movements of the planes in the sky as well as on drive ways inside the circonference of the airport. No failure is admitted onthis power plant so that one of the most important point of this project is its highest level of reliäbility.
The total peak power of the photovoltaic array field is fixed to 50 kWp in order to supply the totality of the 5 kVA needed by the user all along the year, during minimum 6 months.
A battery is used to assure the autonomy of the system and to fix the input voltage of 5 kVA PWM inverter to make the DC/AC conversion.
A separate monitoring system will collect meteorological and energetic data so that to notice the operating of the power plant and calculate its global efficiency.

I. INTRODUCTION

The experimental photovltaic power staion of NICE airport is one of the pilot units of the photovoltaic scheme launched in 1980 by the European Communities.

This is one of the two installations settled by PHOTOWATT on the French territory within this scheme, with :

- ALES MONT BOUQUET Power station.

PHOTOWATT INTERNATIONAL S.A. supervised the complete implementation of this station, as well as of the station of Mont Bouquet.

II. GENERALS

The scope of the project is to use photovoltaic solar energy produced by the station to power the electronic devices of the control tower and ground signalling to aircrafts.

Those sophisticated apparatuses such as goniometer vectors, image number analysts or sound measurement computers, require a supply of electrical energy which is free of distorsion and power micro cut.

EDF system, although it is backed by generators, could not avoid some disturbances, more particularly during stormy periods, while switching from Normal to Emergency.

A photovoltaic power station is the solution to such problems because in its application, it is independent from the main.

The area used to settle 4 rows of photovoltaic captors is situated in front of the freight terminal, and forms a rectangle of 150 metres long and 20 metres wide, oriented South.

The so-formed photovoltaic field is divided into 70 branches of ten 72 peak-Watt modules, cabled in series.

The last branch, cabled in 48 Volts, serves to power all the measuring, display and data acquisitions sections of the station.

Each of the 70 branches is connected to a battery charge regulator, monitoring the disconnection of excess photovoltaic branches, in case of battery over-charge. Regulation of the device is operating according to the voltage and temperature of the battery elements.

Total power consumed by the different devices fluctuates between 5 and 6 KVA permanently, with an annual consumption of 48000 KWh.

Allowing a complete autonomy of a station of such power would have required a very important accumulators bank for inter-seasonal storage.

The setting of a low power charger helped to balance the loss of charge during winter time while reducing the size of storage.
Hence, the photovoltaic field of 50 KWp will supply, according to meteorological conditions prevailing in the city of NICE, approximately 48000 KWh per year, giving a net available energy of 38000 KWh, considering the yields of the different equipments, namely the charge regulator, the battery, and the converter.

Consequently, total net share of photovoltaic energy will average approximately 80 % of the year.

III. STATION CONFIGURATION

The part "captation of solar energy" is made out of 710 modules of 72 peak Watts (mean value), and shows an active area of 700 m2, mounted on the terrasse of the airport.

Total peak power of the field reaches 50 KW, according to standard measurement conditions.

The battery is made out of 106 elements of 1500 Amps-hours, in series, assuring a nominal voltage of 212 Volts.

The protection cabinet includes all the circuit cutting elements able to isolate each of the sub-systems in case of failure or incident, and allow intervention with all security.

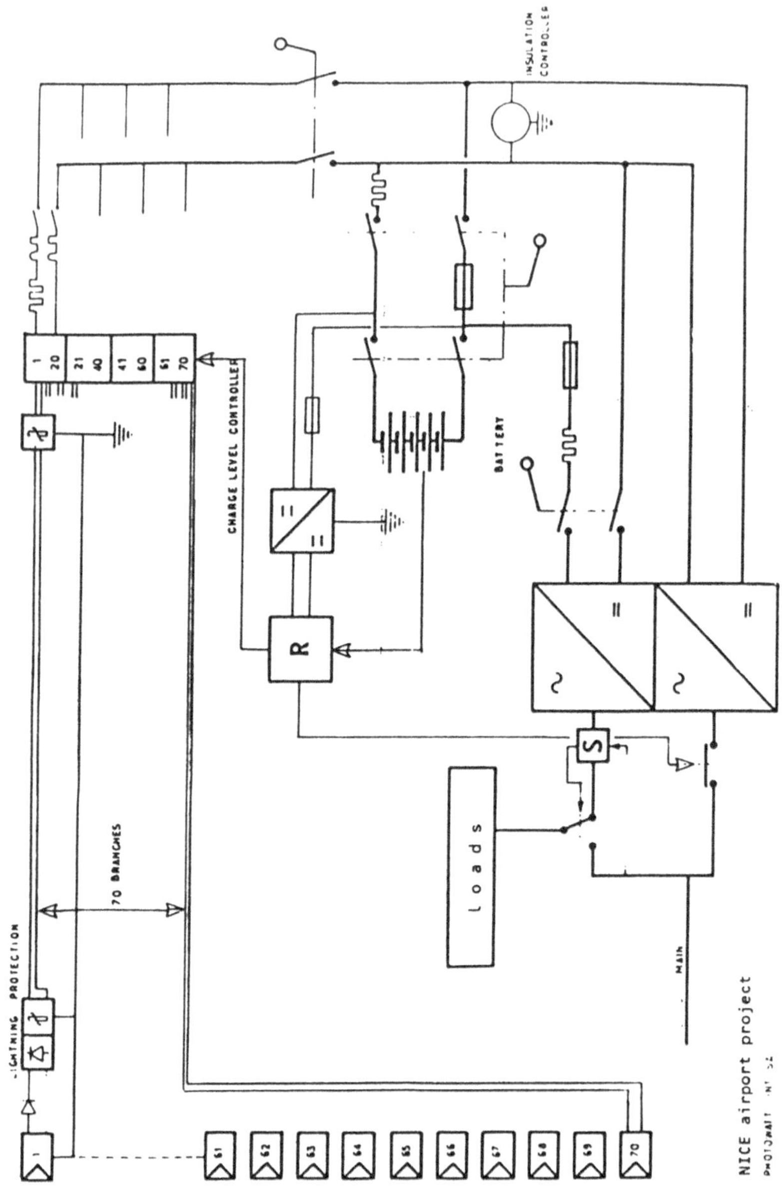

NICE airport project

Conversion of electric direct current to alternating current is made thanks to an electronical converter (chopping type), delivering a monophase 220 Volts alternating current, able to power the different control equipments.

Considering the comparatively small size of the accumulator bank, a charger will reliveve the photovoltaic field in case of long periods of bad weather conditons.

A certain number of steps were taken in order to guarantee the life-span of the different electronic equipments and modules, in case of lightning strikes nearby the photovoltaic field.

The whole unit is built according to prevailing electrical standards, and includes all the requested protection as, especially, an isolation control unit to watch the whole installation.

IV. WORKING PRINCIPLE OF THE STATION

The charge due to control equipments has constant consumption of 6 KVA all year long.

Hence, the profile of consumption may be represented as follows :

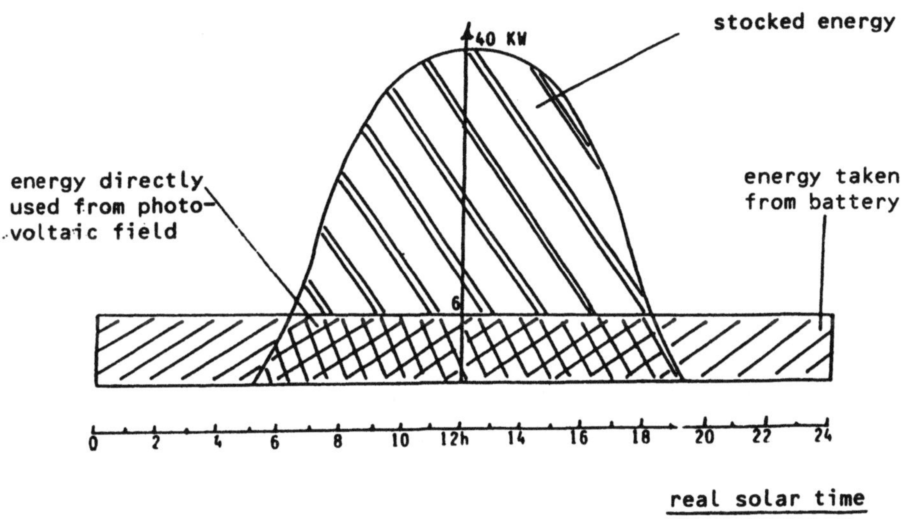

Maximum available power from the photovoltaic field corresponds to some 40 KW. Practically, this value is only reached at noon (solar time) in very favourable conditions.

For that reason, batteries can stock excess energy during the day, in order to restitute it every night.

Ideally, batteries would have had to be sized in order to stock excess energy of favourable summer period, to compensate for winter deficit period. But this would have led to a sizing about ten times bigger than the chosen battery, and consequently to a considerable cost for installation and maintenance, which is apart from future projects for countries where the sun is more regular over the year.

Considerations of above reasons have oriented the choice to a battery charger of power nearly equal to consumption. Hence, solar energy will recharge the battery during periods without sun, while the charger will take care of consumptions.

It works in parallel with the photovoltaic field, and does not have its own regulation. it is governed, hit-or-miss, by the regulation system of the photovoltaic field. Its contribution is 35 % in very unfavourable periods and becomes nil for more than six months a year.

V. DESCRIPTION OF SUB-SYSTEMS

5.1. Modules

- made by PHOTOWATT INTERNATIONAL S.A.
- Type PW P-800
- peak power 73 W
- voltage 12 or 24 V
- laminated technology (glass/PVB/tedlar[R]/alu/tedlar[R])
- 72 monocrystalline silicon cells, Ø 100 mm
- metallic connection box stuck at rear of module
- wire connection with "fast on" plugs
- protection against "hot spot" phenomena by 6 diodes settled in connection box
- self-bearing aluminium frame
- installation with 4 bolts, from front face of module

5.2. Metallic structures

Each branch of the photovoltaic field is made of 2 panels of 5 modules.

While developing light and adaptable structures suitable for different configurations of panels and various tilt angles, a certain number of technico-economical conditions were taken into account :

- no redundant part between frames and modules
- no necessity to bore holes

- no elctrochemical pile phenomenon
- little manpower needs for erection.

As a result, the metallic frame uses the frame of the module as part of the structure itself.

The unit is principally made of 2 high and low girders in extruded aluminium. Their profile has been especially studied to allow the setting of assembling and cabling accessories without boring, and the setting of a rear "pi"-like profile, studied to abide forces created by winds up to 180 km/h. The whole structure is mounted upon 4 pipes in galvanized steel, anchored into prefabricated reinforced concrete girders through shell-like aluminium articulation pieces, allowing perfect ajustment of height and tilt angle without need to make precise fourndations.

These structures were designed and tested according to French standards for buildings construction.

5.3. Cabling

The nominal working voltage is set at 220 Volts, direct current, to take into account unavoidable charge losses in cables, due to long sections of cable used in the photovoltaic field.

That length varies from 30 to 150 metres. The high tension means that all metallic parts must be linked to the ground. In this respect, a ground wire interconnects the frames and the connection boxes, and is also connected to the main of the building.

This conducting wire also serves to evacuate currents induced into the cables by lightnings.

The 71 cables connecting the branches of the photovoltaic field to the battery charge regulator are settled in galvanized steel troughs onto the frames with adjustment parts bolted on low girders of panels.

5.4. Lightning protection

Protectiong against direct lightning strikes is most difficult with such a surface, and the essential step taken was a careful linkage of the different metallic parts to the ground.

General protection of the entire airport must normally avoid the arising of such problem. on the other hand, currents indirectly induced into the cables following lightnings are an important risk.

For that reason, a specific modular protection, adapted to the problem, was developed by "SOULE EM" Company.

The basic principle is not to set any polarity compared to the ground, but to limit over-voltages between phases and ground, and between phases, with 3 adapted varistances, located at both ends of each cable.

An additional protection, on the module side, allows to protect the diodes in the connexion boxes of modules.

Choosing and sorting the varistance is of utmost importance, considering the number of modules in question, and the open-circuit voltage, which overpasses 400 Volts.

5.5. Charge regulator with sequential monitoring

The essential aim of a charge regulator is to allow the stocking of energy in batteries, in such a way as to preserve their life-span.

Hence, it has to limit the in-coming current in the battery, while maintaining voltage below a pre-determined threshold, which itself evoluates according to temperature. Any over-passing of that voltage would cause damage to the plates and an over-consumption of water.

Different solutions are available to fulfill this function, and PHOTOWATT INTERNATIONAL S.A. specifically developped an original cutting system to palliate drawbacks of classical systems.

The principle relies on the comparison, every thirty seconds, of the working point of the battery with a pre-set temperature compensated data. Result of the comparison induces the putting ON or OFF of one of the 70 branches of the photovoltaic field.

The maximum current increase or decrease is thus only 2,5 Amp, and it is not necessary to provide for the hysterisis phenomenon.

On that account, battery charging is achieved in optimal conditions. Commuting parts are made of special relays placed in series on each branch of the photovoltaic field. Sequence monitoring is made through logic cards assuring an overall consumption of about 30 permanent Watts.

5.6. Accumulators bank

The cycling conditions prescribed to the battery should not exceed 25 % in order to guarantee a working life-span of 8 to 10 years. Choice was set on OLDHAM lead elements showing little contents of antimony, and having the following characteristics :

- 106 elements in series, SZT12 type
- capacity 1500 Ah at 10 hours discharge rate
- low antimony content
- tubular positive plates, with oxide added
- nominal voltage : 212 V
- low self discharge rate
- weight : 10 tons

5.7. Conversion DC/AC

The direct/alternating conversion is achieved thanks to a high performance converter, derived from the standard range developped by "AEROSPATIALE" Company.

Indeed it was necessary to take into account the direct input voltage delivered by the installation.

The main features are :

- power 5 KVA
- monophase autonomous inverter, 220 Volts
- chopping with impulse width modulation transistors
- input voltage from 180 to 260 V, direct current
- output voltage tolerances 220 V $\pm$ 1 %
  - . distorsion less than 4 % with non-deforming charge
  - . frequency 50 HZ $\pm$ 1 %

  within the whole range of input voltage and charge power from 0 to $P_{nominal}$
- curent limitation 1,25 $I_{nominal}$ during t from 0 to 100 seconds
- yeld of inverter with cos φ = 1 higher than 90 % between 40 % and 100 % P_n.

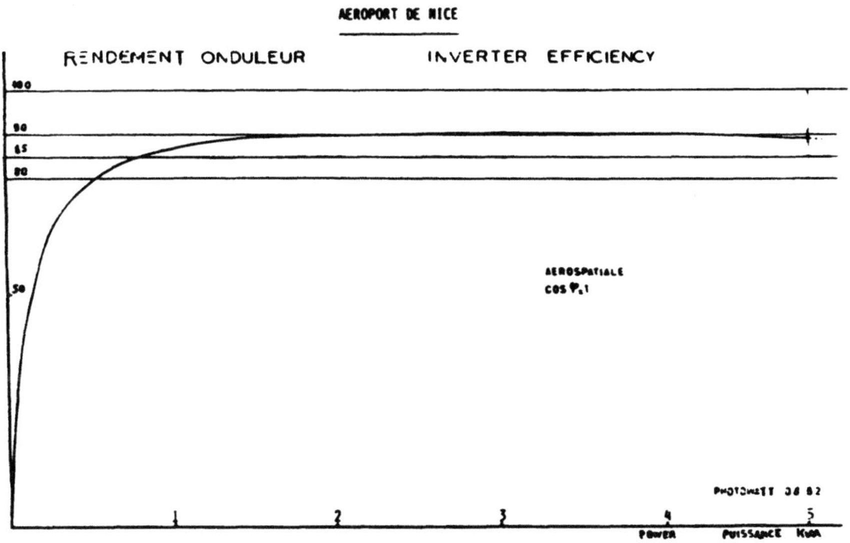

5.8. Charger

The alternating/direct charger is a standard product of a series developed by AEROSPATIALE, and is part of the same cabinet as the inverter.

This charger works in parallel with the photovoltaic field during periods without sun.

The regulation is assured by the solar system itself.

The main features are :

- power 7 KVA
- input voltage three-phase 220 - 380 V
- output voltage, direct current, 180 and 260 V.

The monitoring in the same cabinet as the inverter led to an automatic commuting system on the main, in case of undesired failing of the inverter, so that the efficiency will not be altered.

5.9. Linkage with low voltage board

Distances between the technical premises of the photovoltaic station and the two delivery points are approximately 300 and 600 metres.

Low voltage cables must therefore run through technical galleries and existing sleeves up to the delivery points.

6. CIVIL WORKS

The freightterminal was erected in 1953 and has an inaccessible terrasse which did not originally take into account the weight of photovoltaic equipments.

Checking its structure led to determine the limits of resistance and firmness of the building.

From this study came out an original method of setting solar panels.

Panels are fixed to reinforced concrete elements which are simply laid on the terrasse. Their weight was calculated so that the structure of the panel would resist to overturning efforts due to winds.

Each prefabricated element weighs about 500 kg, and is put on a 3 cm thick isolating element which ensures an even charge distribution.

In order to avoid tightness problems, reinforcements were added to rows of girders to eleiminate any stamping or deterioration of sealed parts at moment of installing or with time.

The diversity in tilt angles and levels of the terrasse led to developing certain special parts to help lining up the rows of panels.

The set of 284 prefabricated elements weighing 140 tons was elevated to the terrasse with a crane.

VII. ASSEMBLING - TRANSPORTATION - ERECTION

The design of supporting metallic structures allowed the transportation of pre-assembled panels. It was taken advantage of this possibility, since the 142 panels were mechanically assembled in the factory, cabled, and checked with specific mounting kits.

This operation led to a substantial reduction of assembling time on site.

Panels were carried to site by lorry, and settled within nine days by M.G. FRANCE.

VIII. DATA ACQUISITION AND PROCESSING

One of the objectives of European launchers of photovoltaic power station projects is to evaluate the evolution of the performance of each project with time.

Therefore it was necessary to equip each power station with relatively complex measuring instruments, and to deliver information suitable for comparison.

The basic principle leading to the definition of the system is to consider this measuring unit as completely independent from the power section of the station, in other words excluding any active role in its functioning.

The principal aim of the monitoring system is to collect all data regarding energy, meteorology or state of generator, to put those in shape, then to visualize them on a synoptic panel while being tape-recorded. This resulted in the following configuration.

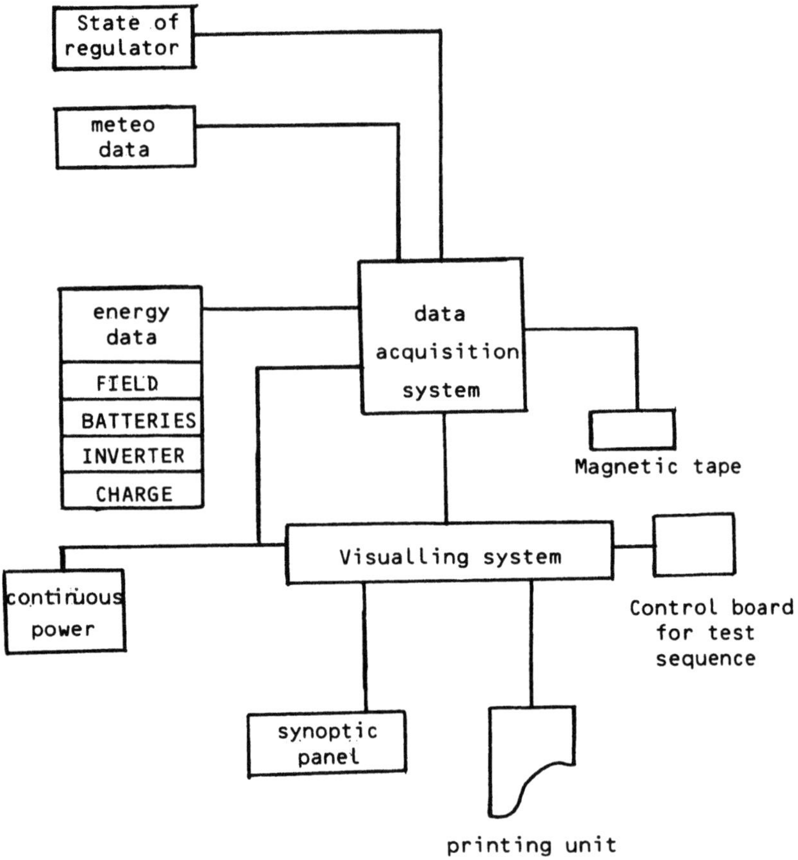

The 71st branch of the photovoltaic field, cabled in 48 Volts, with its independent battery, is an autonomous generator assuring power, without external help, for the entire monitoring system which is, under these conditions, completely separated from the main generator.

Data are measured with isolated captors, and forwarded to a first sub-unit which transforms the information into a digital form, and makes the necessary mathematical calculations to sypply mean values, extremes, or integrals upon given time periods, and which registers these data on a DC 300 XL cassette in a pre-determined form, common to all projects.

These intermediate informations are taken from this first module towards the next one which pilots a visualization board, for instant or cumulated informations, and a printing unit. It also proceeds to testing each photovoltaic branch.

MEASURED PARAMETERS

meteorological	. global horizontal illumination
	. illumination on the panels plan
	. outside temperature
	. wind speed
	. wind direction
	. effort on structure
state of generator	. number of branches in service

ENERGY DATA

field	. voltage in a branch
	. temperature of a cell
	. current output
battery	. voltage
	. current
inverter	. current in
	. current out
charge	. power taken from main

Measuring information is done every second, and registration of results every hour.

IX. WORKS SCHEDULE

The main phases in the realization of NICE AIRPORT power station were as follows :

- study phase started in January 1981
- design and development of components during 1981
- manufacture of materials from January to September 1982
- pre-assembling of panels in factory in November and December 1982
- realization of girders and preparation in October 1982
- craning of girders in November 1982
- delivery to site of power parts in January and February 1983
- mounting, cabling of panels in February 1983
- starting up of power part in March 1983

- charging of equipments in March 1983
- study, development, realization and starting up of monitoring system from January 1982 to AUGUST 1983
- reception SEPTEMBER 1983

X. ECONOMICAL ASPECTS

The installation of a power station of the size of NICE Airport allows us to evaluate the respective weight, cost wise, of the different parts of the unit, as well as to split the respective parts of equipment and of research, and development works.

This is shown in the following chart :

	TOTAL	material	R ¤ D
Modules	41 %	100 %	0
Frames / cabling	3 %	80 %	20 %
Energy transfer	3 %	90 %	10 %
Regulation / protection	6 %	70 %	30 %
Batteries	4 %	95 %	5 %
Conversion DC to AC	8 %	90 %	10 %
Civil works	13 %	95 %	5 %
Transportation / erection	10 %	80 %	20 %
Monitoring	5 %	60 %	40 %
Miscellaneous	7 %	60 %	40 %
	100 %		

Hence, it appears that, apart from the modules part which remains an important one, major parts pertain to civil works and transportation/erection, representing nearly a quarter of the total.

This point particularly stresses the importance of the choice of structures, their means of anchoring to the ground, and the possibility to pre-assemble the panels in factory which limits interventions on site.

THE FOTAVOLTAIC PROJECT - A 50kW PHOTOVOLTAIC SYSTEM TO POWER A DAIRY FARM ON FOTA ISLAND, CORK, IRELAND.

Contract number: ESC-P-061-EIR(H)

Duration : 28 months 1 March 1981 - 30 June 1983

Site of the installation: Fota Island, Cork.

Power: 50 kW peak

Total Budget : IR£737,230

Sponsors beside the CEC: BMFT (D)
AEG-TELEFUNKEN (D)
ELECTRICITY SUPPLY BOARD (IRL)
NATIONAL BOARD FOR SCIENCE & TECHNOLOGY (IRL)
DEPARTMENT OF ENERGY (IRL)
UNIVERSITY COLLEGE CORK (IRL)

Head of Project: Professor G.T. Wrixon

Address : National Microelectronics Research Center,
University College,
IRL - Cork.

Members of consortium: University College Cork
AEG-Telefunken
Electricity Supply Board

Summary

The 50kWp photovoltaic installation built at Fota Island is described. Its purpose is to supply power to operate a dairy farm whose seasonal power needs are well matched to the available insolation. 2775 modules, each with a nominal output power of 19.2W and consisting of 20 multicrystalline silicon solar cells (10 x 10 cms) connected in series, comprise the array. These modules are divided up into 13 sub-arrays each consisting of 6 parallel strings of 36 modules in series, which can deliver 14.5A at 268V. The sub-arrays are individually switchable to one of two DC bus-bars each connected to a battery of capacity 300Ah at 268V. The dairy farm load, which mainly consists of motors, has been divided into 3 parallel circuits each driven by a 10kVA high efficiency self commutating inverter. The loads are fed, via the inverters, from one battery bus while the other battery bus is charged by the solar array. Switching of the sub-arrays, inverters, loads etc is under the control of a microcomputer which will assure optimum operating efficiency of the installation. A 50kVA line commutated inverter is used to export surplus energy to the utility grid and to provide backup during periods of prolonged low insolation. The modules are mounted in one plane at 45° elevation in 25 rows of 111 modules each. The back and sides of the array are enclosed by a concrete block structure to reduce wind loading.

1. Introduction:

The Fotavoltaic project involves the supply of electrical power to a dairy farm, milking 250 cattle, using a 50kWp solar generator independent of the utility grid. The solar array consists of 2775 solar modules mounted on the roof of a south facing building on the dairy farm. The roof is inclined at an angle of 45° to the horizontal. At the centre of the solar building is the control room and a battery room. Power conditioning is performed in the control room and the power is then delivered to the dairy via a 200m underground cable. Excess energy from the system is delivered to the utility grid using a line commutated inverter. The utility grid is also used to recharge the battery when it becomes deep discharged. A microcomputer located in the control room controls the system's operations. As well as controlling the system it monitors data on a computer terminal, records data on magentic tape and protects the system from faults. Peak load demand of 90kWh/day exists during the summer months and this decreases to less than 5kWh/day in mid-winter. Solar array output during the summer months exceeds 200kWh/day and decreases to 30kWh/day in mid-winter.

2. Technical Description of System:

The system is divided into seven main sections as shown in Figure 1.

(1) Solar Array:

For monitoring and control purposes the array is divided into thirteen sub-arrays as shown in Figure 2, the DC power distribution circuit. Power from each sub-array is switched to either battery 1 or battery 2. The number of sub-arrays connected to battery 1 depends on the level of solar insolation. The remaining sub-arrays are then connected to battery 2.

(2) Battery:

The battery is located in the battery room at the rear of the building and consists of 268 individual battery cells. The battery is divided into two groups each having a nominal voltage of 268V and a capacity of 300 Ah. Normal operation involves connecting loads to battery 1 while battery 2 operates in a stand-by mode.

(3) Loads:

The dairy farm loads consist of milking machine motors, cooling motors and washing motors. To prevent overloading of the inverters and also to ensure high efficiency operation of the inverters the load is divided into three separate groups, each group supplied by an individual inverter.

(4) Utility Grid:

A three phase AC grid connection exists in the control room. The utility grid accepts excess energy from the system and also delivers energy to the system when the batteries become deep discharged.

(5) Inverters:

Two types of inverters are used in the system, i.e. self-commutated

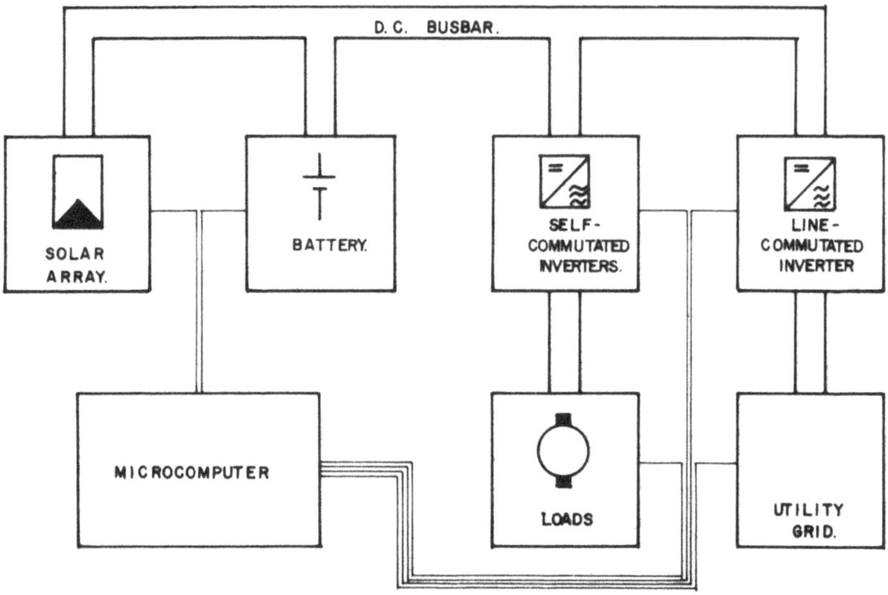

FIGURE I. LAYOUT OF FOTAVOLTAIC SYSTEM.

and line commutated. Three 10KVA self-commutated are used to supply the three phase dairy farm loads. As shown in Figure 3, in the AC power distribution system each load group is connected to an inverter. When a fault occurs on an inverter the load group is reconnected to the utility grid. For safety purposes the two switches which connect the loads to either the utility grid or to the photovoltaic system are mechanically interconnected. A 50KVA line commutated inverter is used to deliver excess energy to the utility grid. This unit can also act as a rectifier and is used to recharge the battery when necessary. The current flow through this inverter is regulated by a 0-10V signal from the microcomputer interface.

(7) Microcomputer:

The microcomputer is used to control the system by controlling each individual hardware unit and each interconnecting switch. As well as controlling the system it ensures that the system operates at its highest efficiency while checking that each unit is protected from faults. The microcomputer consists of a central microprocessor system, computer terminal, printer and magnetic tape recorder.

3. Link to Application:

Powering a dairy farm using a solar generator was selected because as well as dairying being an important Irish industry the energy demand of the dairy farm has the same seasonal variation as the output of a solar generator. During the summer months peak electrical demand occurs on the dairy farm due to increased milk yield and during winter months electrical

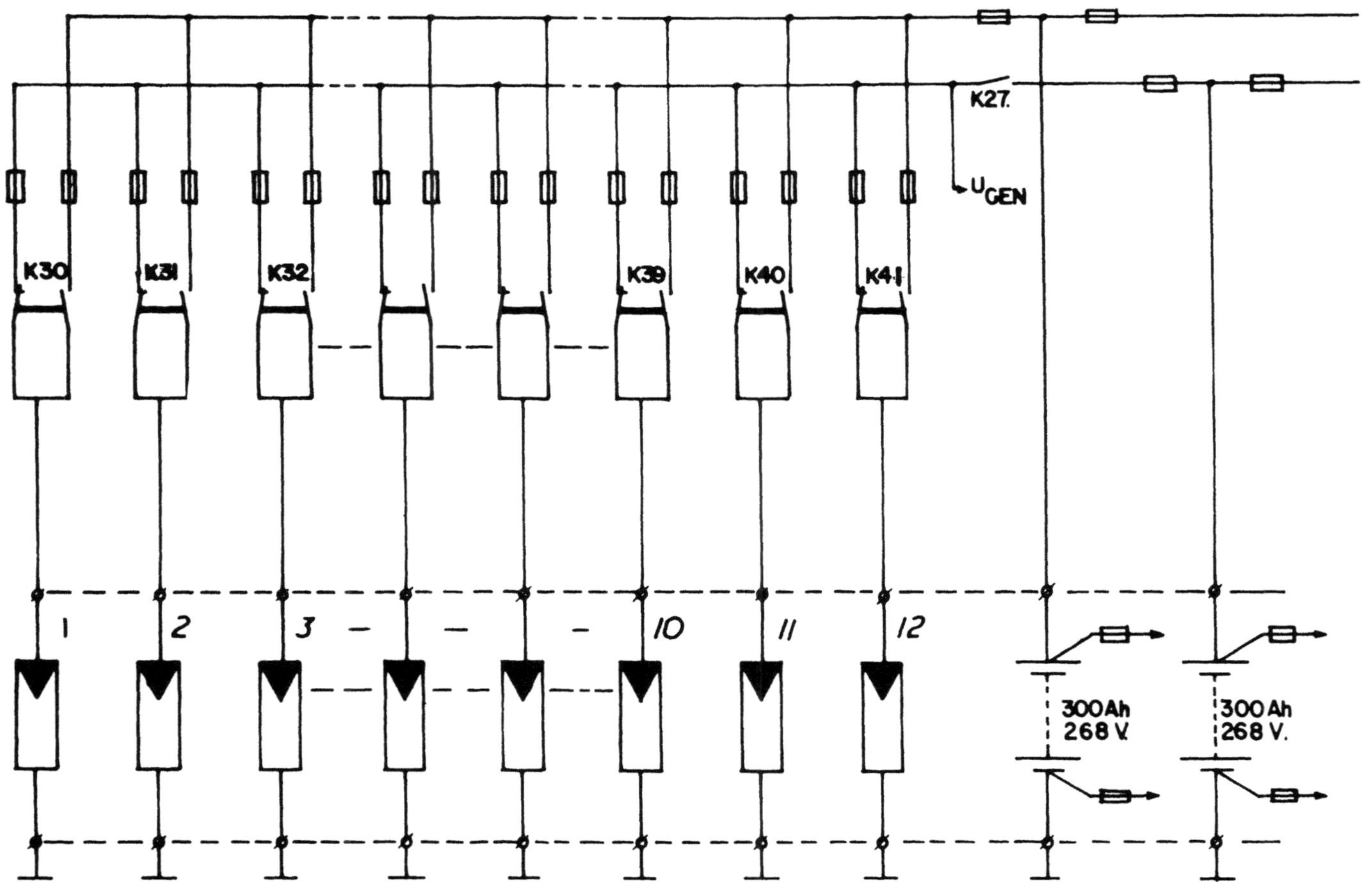

Figure 2: DC Power Distribution Circuit

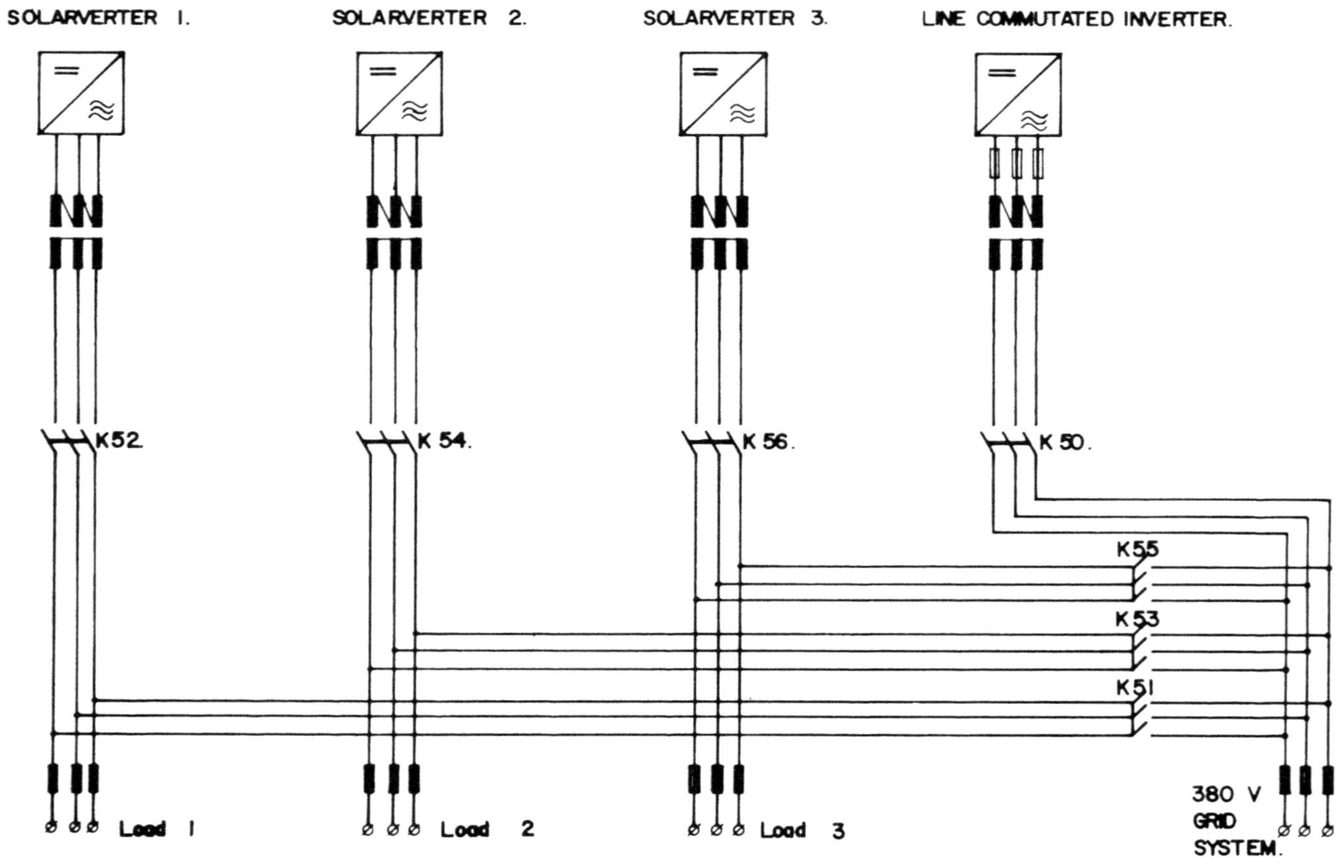

Figure 3: A.C. POWER DISDRIBUTION.

demand decreases due to decreased milk yield and also due to calving. During mid-winter the milk yield decreases to zero.

The electrical load demand on the dairy farm is well defined. Milking occurs twice per day, i.e. in the morning, 7.00 a.m. to 9.30 a.m. and in the evening 4.00 p.m. to 6.30 p.m. During this time maximum demand is 10kW. The main electrical loads are milking machine motors, milk cooling motors, washing motors and feeding motors. When milking is not in progress the only electrical load is the 4.5kW milk cooler which operates intermittantly to cool the 8000 litre bulk milk tank.

4. Architectural and Environmental Aspects:

The solar building is 54m long, 15m high and 15m wide. It is built using 20,000 high density 10 newton blocks. The building is sectioned into eleven bays and the ten internal walls are used to support a galvanised steel structure to which the modules are attached. The roof is inclined at 45° to the horizontal and the building is south facing. The centre bay is used as a control room and battery room, the end bays are used for storage and the remaining eight bays are used as calving units by the dairy farm. The modules are bolted on the steel structure and are therefore in one continuous plane.

Environmental data is available from Roches Point meteorological station seven kilometers from the solar array site.

Temperature:

- Maximum recorded temperature	:	(1876 - 1980)	28.9^{o}C
- Minimum recorded temperature	:	(1876 - 1980)	-6.7^{o}C
- Mean daily maximum temperature	:	(1951 - 1980)	18.1^{o}C
- Mean daily minimum temperature	:	(1951 - 1980)	3.9^{o}C

Wind Speeds:

- Maximum recorded wind speed	:	(1963 - 1979)	32.38m/s
- Predominant wind direction	:	South westerly	

Cleaning of the modules is not necessary due to regular rainfall. Two test modules were mounted at the solar array site in October 1980 and observations indicate that manual cleaning is not necessary.

5. Overview of New Technological Developments:

The development of a microcomputer for the control of the photovoltaic system involved three main technological developments. Firstly, the development of a photovoltaic hardware system which could be interfaced to the microcomputer. Secondly, using this hardware a system control switching was developed and thirdly a software package was developed which implemented this control.

The microcomputer consists of a Motorola 6800 microprocessor system, computer terminal, printer and magnetic tape recorder. 24K of software is stored in EPROM + 8K RAM. Analog inputs to the microcomputer interface from the PV system are represented by 0-20mA signals and digital inputs representing status of switches are represented as 0, 24V digital signals. Internal calculations are performed using these inputs and switching arrangements are implemented using 0, 24V digital output signals.

The microcomputer has complete control over each hardware unit and each interconnecting switch in the system. By defining the system's operations in terms of switch configurations the correct energy flow in the system is controlled. The main control operations performed by the microcomputer are:

(1) Battery charging using solar array,
(2) Power supply to dairy farm loads,
(3) Delivery of excess energy to utility grid,
(4) Battery charging from utility grid.

As well as controlling these functions the microcomputer ensures that the correct switching sequence occurs so that each hardware unit is protected in the event of a fault occurring. It also monitors and displays measured and calculated data on a computer terminal and records data on magnetic tape for subsequent analysis.

The microcomputer controls the system automatically but a facility exists whereby operating parameters can be changed manually via a computer terminal. These parameters are analog limits, filter factors on analog measured data and delays on digital data. Other features for testing include setting of limits, manual simulation of measured data and start-up and shut-down of system from the computer terminal.

Changes or expansion of the system control is performed by changing the software on the EPROM's.

6. Modules:

The solar modules are of type PQ10/20/01 supplied by AEG-Telefunken. Each module consists of 20 multicrystalline solar cells enclosed between two sheets of glass and surrounded by a stainless steel frame. At the base of the module two wires protrude which have water tight connections. Also integrated in each module is a shunt diode. Each module is rated at 19.2Wp with a working voltage of 8.6V. The operating characteristics are as shown in Figure 4.

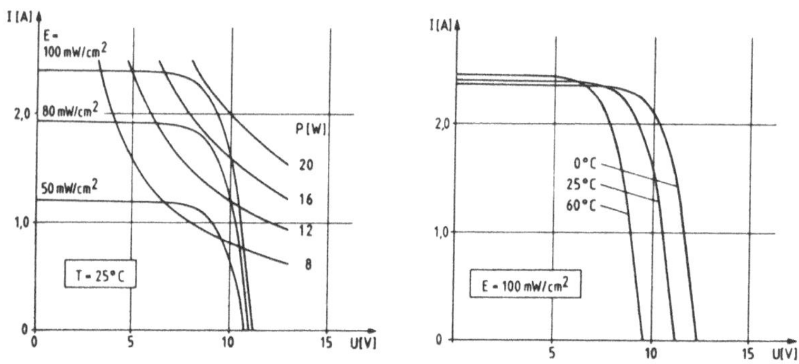

Figure 4: Voltage/current characteristic of solar module PQ10/20/01.

7. Solar Array and Cabling:

The solar array consists of 2775 solar modules mounted on a steel structure. Each module is held in position by two bolts as shown in Figure 5. A torque of 7.5 was applied to each bolt. For monitoring and control the solar array is divided into thirteen sub-arrays. Each sub-array

Figure 5: Module fixing bolt

contains six parallel strings and each string has thirty six modules connected in series. Each sub-array has an open circuit voltage of 403V and a short circuit current of 14.4A.

Each sub-array has a junction box which is mounted on the inner wall of the building as shown in Figure 6. A 2.5mm^2 cable from each string is connected to its junction box and from the thirteen junction boxes a three core 10mm^2 cable is delivered to the control room. Each module's location and power rating is known. When a fault occurs in the array the micro-computer detects a decrease in the sub-array current and prints a message indicating the sub-array. The faulty string is then located by testing the strings in the sub-array junction boxes. Three modules have built-in PT100 temperature sensors and these are connected to the microcomputer interface for monitoring.

8. Array Support Structure:

The module support structures are made from galvanised steel and rest on the internal walls of the building (see section 4). The steel structures were designed by a computer analysis for various load cases. The critical case was for wind suction of 2Km/m^2. This wind loading

Figure 6: Sub-array Junction Box

results from a wind speed of 50m/s and resulted in a calculated maximum deflection of 12.0mm at the centre span. The maximum bending stress on the structure was $120 N/mm^2$. Seventy seven steel structures (Figure 7) are used and they are bolted together using 12mm bolts. The steel is connected to the internal walls using teak wall plates.

9. Lightning Protection:

Two methods of lightning protection are used:

(1) Protection of building and steelwork,
(2) Protection of electrical equipment.

The building and steelwork is protected using seven galvanised steel lightning rods at the top of the building. The rods are connected to the earth grounding system using $8mm^2$ solid galvanised steel. The steelwork is bonded continuously at the top and bottom using $16mm^2$ stranded copper (PVC/PVC). The steelwork is connected to the earth grounding at its four corners to minimise the resistence to earth. All metal ducting on the walls is grounded in the same manner.

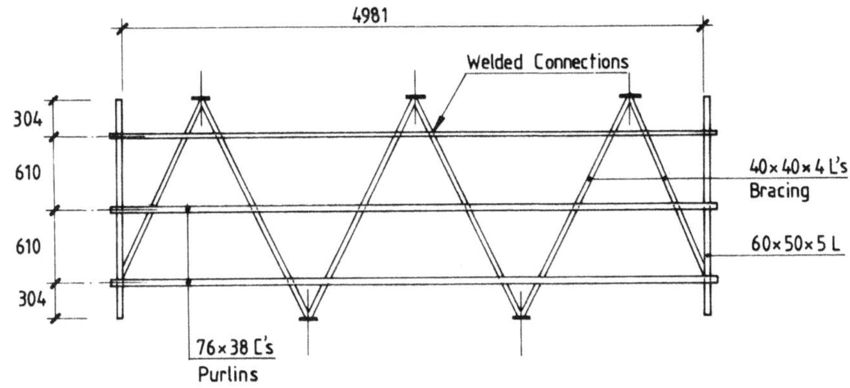

Figure 7: STEEL SUPPORT STRUCTURE USED FOR MODULE MOUNTING

Protection of the electrical equipment is performed using a varistor and spark gap combination in each of the thirteen sub-array junction boxes as shown in Figure 8. When the DC busbar in the junction box exceeds 600V DC the varistor and spark gap acts as a short-circuit to earth thereby protecting the equipment in the control room from large surge currents. The varistors and spark gaps are positioned at the top of the junction box as shown in Figure 6.

10. Power Conditioning and Batteries:

Power conditioning is performed using two types of inverters. Self-commutated inverters are used to supply the three phase AC loads and a line commutated inverter is used to deliver excess energy from the system to the utility grid.

Three 10kVA self-commutated inverters known as 'solarverters' accept DC inputs of 245V-315V and provide an output of 380V three phase AC. Switching transistors are used to provide the output waveform thus eliminating the quenching circuits required for thyristor operated inverters. The solarverters operate at 95% efficiency under full-load and are protected from faults by an immediate shut-down when short circuit or overload occur. When a fault occurs a signal is sent to the micro-computer which then reconnects the load to the utility grid. A message is also printed indicating the fault and the time it occured. As a safety measure the solarverters are manually reset after shut-down.

The 50kVA line commutated inverter known as the mini-semi is a thyristor operated unit. The 50Hz signal is obtained from the utility grid. Consequently, in the absence of the grid the inverter does not operate. The current flow through the mini-semi is regulated using a

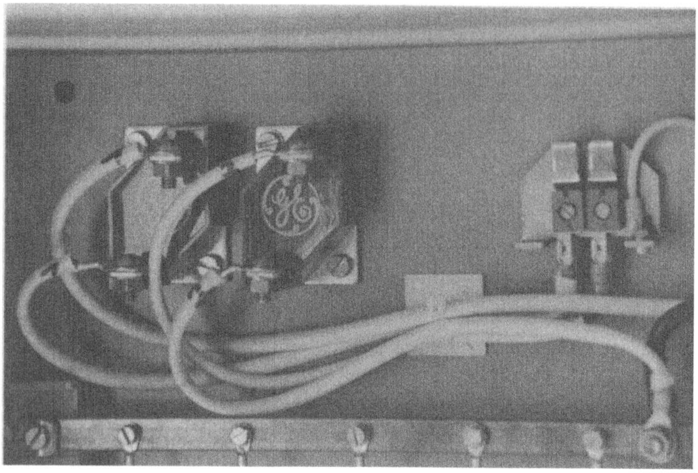

Figure 8: Varistor and spark gap lightning protection of the electrical equipment

0-10V DC signal from the microcomputer. The direction of current flow (inverter or rectifier operation) is also controlled by the microcomputer. When a fault occurs in the mini-semi a signal is sent to the microcomputer. A new switching arrangement is then adopted to protect the system from faults due to excess energy.

Lead acid batteries supplied by Varta are used. Each battery cell has a nominal voltage of 2V and a capacity of 300 Ah. One hundred and thirty four battery cells are then connected in series to form a battery group, see Figure 9. Battery charging using the solar array is regulated by the microcomputer. The charging current is maintained between 50A and 65A by switching the sub-arrays to and from the battery. The number of sub-arrays connected to a battery group depends on the insolation. The remaining sub-arrays are switched to the other battery group which acts in a standby mode. When faults occur in the battery (i.e. deep-discharge, gassing, overcharging, etc) the microcomputer introduces a new switching arrangement to protect the battery and to maintain a proper charge/discharge cycle. Hydrogen and oxygen are recombined by a catalyst in the recombinator thus returning the water to the electrolyte and minimising battery maintenance.

11. System Integration and Installation:

The control room is located at the centre of the solar building. It contains eight cabinets in which are mounted the electrical equipment used to control the energy flow in the system.

Cabinet 1 has the solar array input cables, solar array switches and

Figure 9: General view of battery bank

the current and voltage transducers which deliver analog signals to the microcomputer.

Cabinet 2 contains DC power switches which connect the inverters to the battery groups. It also contains transducers for battery current and battery voltage. Two logistronic units indicate the state of battery charge. The logistronic also sends digital signals to the microcomputer indicating deep discharge, gassing and full charge.

Cabinets 3, 4 and 5 contain the solarverters and output transformers.

Cabinet 6 contains the mini-semi (line commutated inverter), AC power switches and AC transducers.

Cabinet 7 contains utility grid component, i.e. meters and protection equipment. Undervoltage and underfrequency protection is required when exporting from the system to the grid.

Cabinet 8 contains the microcomputer. It consists of twenty plug-in cards mounted on a standard intermas rack system. The connection of input and output signals to the microcomputer interface is done via connecting blocks mounted on the rear of the cabinet.

Figures 10 and 11 show Cabinets 2 and 3 respectively. The cabinets are mounted on a steel frame and interconnections between cabinets is done through this frame.

12. Monitoring System

Monitoring is performed using the same microcomputer which controls the system's operations. A computer terminal is used to display operator request data and is also used to request the display of data on the printer. The printer records messages indicating switch operations

Figure 10: Battery control cabinet including logistronic units

and fault conditions. Error messages are printed indicating faults, the time they occured and the sequence in which they occur. These messages are held as a permanent record of switching arrangements and fault logging. The following example shows the type of message printed.

0553	BATTERY 1	TEMPERATURE 1	56.0^{o}C	**B	8 : 01.03
0553	BATTERY 1	TEMPERATURE 1	54.0^{o}C	E	8 : 05.23

The letter 'B' indicates the time the fault began and the letter 'E' indicates the time the fault ended.

Data is displayed on the computer terminal in a data group. Each data group contains measured and internally calculated data. Twenty data groups exist and examples of these are solar array output data, battery 1 data, battery 2 data, solarverter 1 data, solarverter 2 data, solarverter 3 data, line commutated inverter data, meteorological data, etc. The following example displays the format of the data group indicating the output of the sub-arrays (Table 1).

SYSTEM JOURNAL	FOTAVOLTAIC PROJECT		11 DEC '81 18.16 LT	
02	SOLAR ARRAY OUTPUTS..I, V.			
0400:	ARRAY VOLTAGE..U.GEN.		251.	V.DC .....
0401:	OUTPUT CURRENT OF SUB-ARRAY	1	9.56	A.DC .....
0402:	OUTPUT CURRENT OF SUB-ARRAY	2	9.02	A.DC .....
0403:	OUTPUT CURRENT OF SUB-ARRAY	3	9.56	A.DC .....
0404:	OUTPUT CURRENT OF SUB-ARRAY	4	9.50	A.DC .....
0405:	OUTPUT CURRENT OF SUB-ARRAY	5	9.56	A.DC .....
0406:	OUTPUT CURRENT OF SUB-ARRAY	6	9.16	A.DC .....
0407:	OUTPUT CURRENT OF SUB-ARRAY	7	9.53	A.DC .....
0408:	OUTPUT CURRENT OF SUB-ARRAY	8	9.51	A.DC .....
0409:	OUTPUT CURRENT OF SUB-ARRAY	9	9.03	A.DC .....
0410:	OUTPUT CURRENT OF SUB-ARRAY	10	9.20	A.DC .....
0411:	OUTPUT CURRENT OF SUB-ARRAY	11	9.29	A.DC .....
0412:	OUTPUT CURRENT OF SUB-ARRAY	12	9.42	A.DC .....

Table 1: Printout of data group showing output of sub-arrays

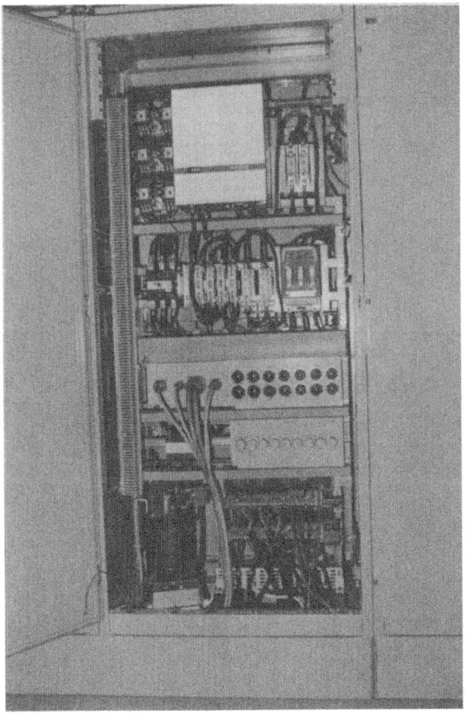

Figure 11: AC control cabinet including the mini-semi line commutated inverter

13. Conclusions and First Results:

The integration and installation of the equipment is completed and tests have been performed on each unit. Testing of the computer software is still in progress and changes and improvements in the system's operations are still being made. The microcomputer facility for allowing changes in delays, limits, filter factors, etc. enables an optimisation of the rate of charging, the rate of switching, the rate of measurements, etc. These changes are made via the computer terminal and when satisfactory parameter values are selected these are then burnt into EPROMs.

Data monitoring and data recording is operating well. Initial changes involved adjustment of scaling factors and reorganisation of data displayed in data groups. The first tapes for processing will be available later this year.

Array outputs indicate comparable outputs from each sub-array. However as clouds pass over the array large variations (up to 5A) are observed on the outputs of each sub-array.

Module temperatures of 36°C were recorded while ambient temperature in the shade was 22°C. The three modules with temperature sensors indicate a higher temperature in the upper modules of the array. This temperature difference has a maximum of 3°C.

Initial battery charging was performed using the solar array. The rate of charging safety limits were tested and the switching operations reacted as expected. Due to the large number of switching combinations possible careful observation of switching will continue continuously over the first few months operation.

Figure 12: Cows grazing close to the PV generator

The dairy is operating satisfactorily from the system. Initial problems with the motor's operation were solved with power factor correcting capacitors. Delivery of excess energy to the utility grid is in operation. Two meters were installed to measure energy to and energy from the grid. A maximum demand meter indicates the maximum power to the grid over a two month period.

The work over the next period will involve closely monitoring the operation of the system, optimizing its characteristics via the switching strategy of the microcomputer and maintaining a continuous check on the individual hardware units.

SOLAR-WIND PROJECT TERSCHELLING

Location, country	: Terschelling Island, the Netherlands
Rated peak power of array	: 50 kW
Proposal number	: 015/NL + D/20
Contract number	: ESC-P-068-81-NL(H)
Duration of contract	: 28 months
Total budget of project	: CEC contribution (31%) 1,336,000.-- ECU + 420,000.-- ECU (bonus)
Name of other funding bodies and their share of total costs	: BMFT W.-Germany: 372,000.-- ECU (28%) Ministry of Economics, Holland 225,000.-- ECU (17%)
Contractor	: Main-contractor: Holec N.V. Sub-contractor: AEG-TELEFUNKEN
Project Leader	: Ir. R.P.M. Sonneville Holecsol Systems B.V. P.O. Box 3049 5700 JC Helmond Holland Tel.: 04920-32705 Telex: 59030 hosol nl

Summary

The system provides about 95% of the electrical energy of a school for maritime studies. Local weather-circumstances (irradiation and wind) are used to its optimum by applying a photovoltaic solar generator of 50 kWp in combination with a wind generator of 40 kW. The system is connected to the utility-grid to investigate the interaction and to guarantee a completely reliable energy-delivery to the consumer, secondly to increase the use-factor of the system.

1. Introduction

The Terschelling project will prove feasibility of photovoltaic systems, particularly in the northern part of Europe. Specifically, this project is meant to study the combination of solar and wind energy systems well-matched with energy-demand as a function of the period of the year. The consumer's loadprofile asks for seasonal storage. As the solution of an accumulator bank is absolutely unfeasible, a more practical possibility was considered. Instead of this huge accumulator bank a wind generator of 40 kW will cover energy demand during the winter period. A continuous connection with the public grid is planned.

All relevant techniques needed to control the above-mentioned energy flows i.e. solar, wind, accumulator bank and grid, justify the application of a suitable energy management system. In fact a relative small accumulator bank is installed in this application, because battery charging should be studied, which offers at the same time the possibility to study the complete system in the stand-alone mode.

As exposed in the one-line diagram, the major subsystems are: (see fig. 1)

- photovoltaic generator 50 kWp
- windgenerator 40 kW
- accumulator bank 360 V - 500 Ah
- energy management system
- synchronisable self commutating inverter 50 kVA - 380/220 V
- grid connection equipment.

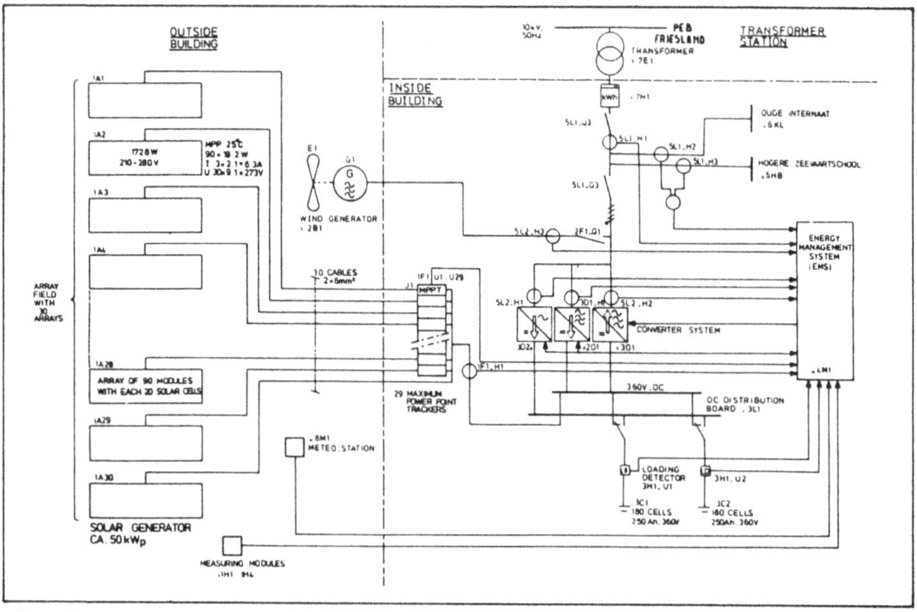

Fig. 1

Furthermore the solar-wind system has been designed to generate sufficient energy to cover the schools yearly needs (35.000 kWh), the remaining (45.000 kWh) will be delivered to the grid. This has the advantage that operational cost can be fully covered.

From the diagramme (fig. 2) can be learned that the load profile is not very well matched with the p.v. output. With the objective to build a self-supporting energy system (stand-alone mode) the combination with a suitable windgenerator is obvious under these typical local conditions.

Solar/Wind System

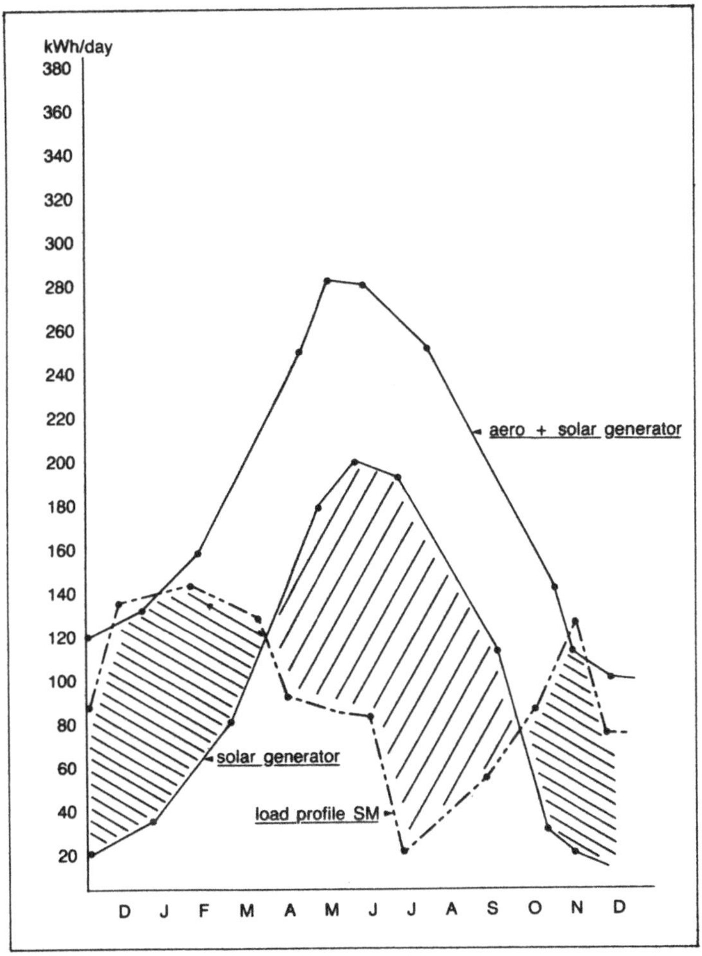

Fig. 2 - Load versus output

2. Environment

Terschelling island is the second largest Wadden-island in the northern part of Holland. It is one of the most well-known Dutch holiday centers. On the island the "Hogere Zeevaartschool Willem Barentsz" is an institute educating in maritime studies. Besides a more than 20% higher insolation, wind-energy availability (6 m/sec annual average) is more favourable compared with data for sites inside the country. Advantage was taken from this situation to get an energy-supply covering the energy demand almost the whole year around.

It must be emphasized that availability of technological skills (teaching staff) in this institute had a rather strong influence during the selection procedure. The school management's interest and their skills assure sufficiently the continuity of the project until 5 years after system-acceptance tests.

The site is more or less predestinated for solar energy. The area in front of the school has a slope of about 10 degrees and is directed to the south. These circumstances offered an excellent opportunity to adapt the solar generator into the landscape. Taking into account the special status of the island and the site in particular, local and provincial authorities had no objections against the way the solar- and windgenerator were planned.

Fig. 3 illustrates the actual situation.

Fig. 3

Fig. 3a - The Windturbine

3. Description of the project

3.1 General

The project consists of a p.v. generator of 50 kWpeak in combination with an accumulator bank as energy storage. In order to avoid seasonal storage, the project is completed with a windgenerator. (Fig. 3)

The solar-generator needs a sand surface of 600 m^2 and is divided into 30 arrays; 29 arrays each consisting of 3 parallel strings of 30 modules and 1 array with 3 parallel strings of each 46 modules. The 29 arrays are equiped with a Maximum Power Point Tracker, providing that under all circumstances power output per array is optimized. The 30th array will be directly switched on the dc-busbar without intervenience of a MPPT.

The installation will be usually grid-connected, the possibility to test the equipment in the stand-alone mode is foreseen. The plant offers also the possibility to connect both the solar- and windgenerator to the utility-grid. In the framework of the P.V. pilot CEC programme this is unique.

As mentioned before the solar and wind energy output will be more than sufficient on a yearly basis. Due to the fact that incidentally power output and power demand are not matched, electrical storage as well as grid-connection is possible.

In order to control the different energy-flows, the conditions for energy exchange to the grid, the electrical maintenance of the accumulator bank and to monitor systems behaviour, an integrated Energy Management/Data-Monitoring System (EMS) is installed.

Except this EMS, and the solar- and windgenerator, the subsystems are installed in the battery- and control room (fig. 4).

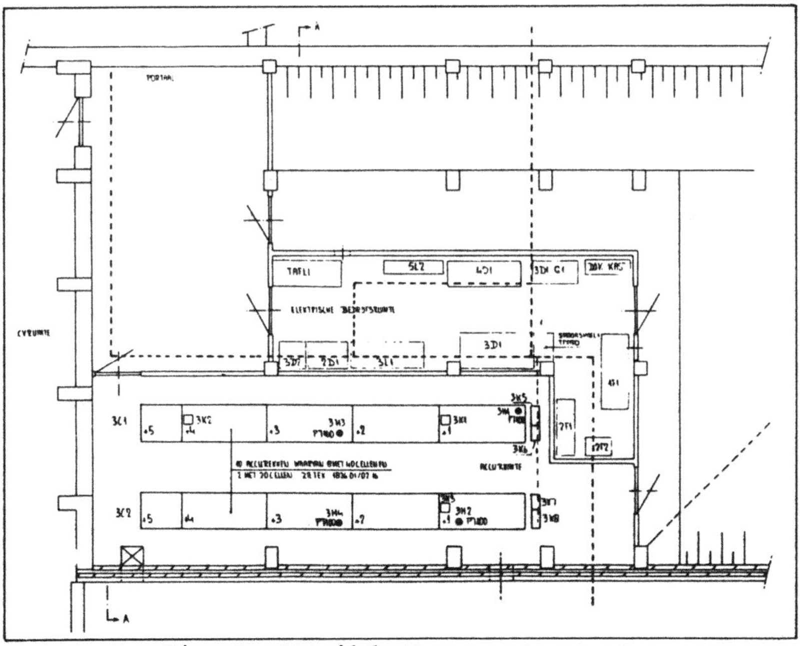

Fig. 4 - Detail battery- and control room

Summarized, the solar-windgenerator system consists mainly of:
- 29 arrays each 90 solar modules
- 29 Maximum Power Point Trackers for power-conditioning and adaption to the dc grid
- 1 array of 138 solar modules
- 1 coupling (reverse) diode for coupling to the dc grid
- 1 dc distribution board
- 2 accumulator banks each 180 cells
- 1 charging rectifier
- 1 reconditioning rectifier
- 1 control cabinet
- 1 Energy Management/Data Monitoring System
- 1 windgenerator 40 kW.

3.2 New technological developments

3.2.1 Maximum power point tracking

In fig. 5 a voltage/current curve is drawn together with the corresponding value of the power. As shown in this figure, maximum power will be generated when current I_m flows in conjunction with voltage V_m (maximum power point).

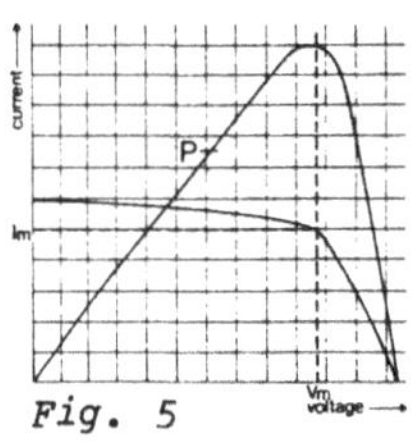

solar-module characteristic

Fig. 5

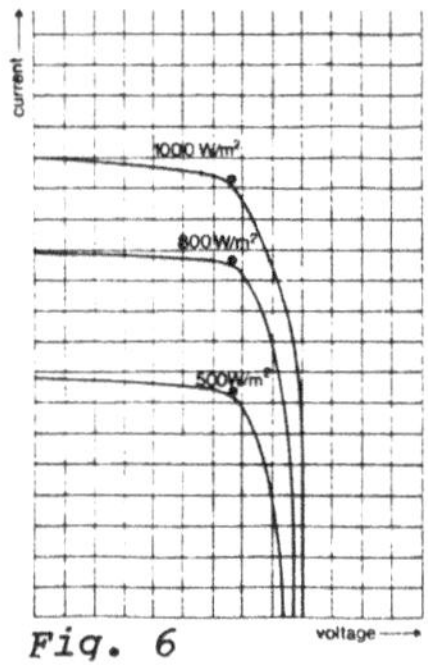

Fig. 6

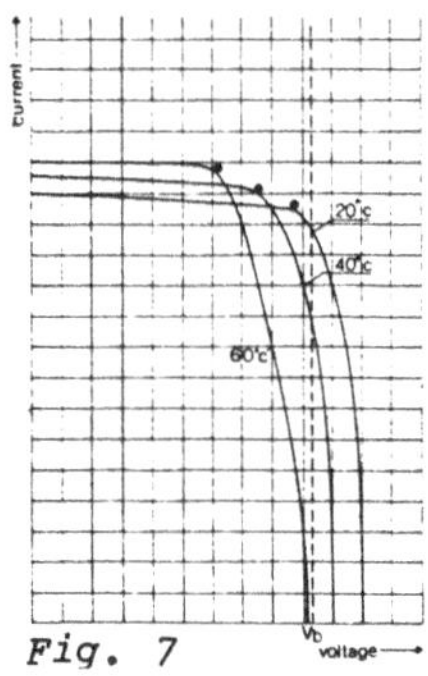

Fig. 7

The maximum power point of a module is not fixed, but will be affected by two major factors:
- the irradiance
- the module temperature.

Due to irradiance fluctuation, I_m will change considerably and a minor change of V_m will occur as well (fig. 6).

More important, however, is the influence of the module temperature. As depicted in figure 7, V_m will decrease with an increasing module temperature causing the module to operate far beyond its maximum power point.

In this figure the battery voltage is assumed to be constant. In practice, however, the situation is even worse: the battery voltage will increase during the daytime, when the battery is being charged.

Consequently, in the maxium power point tracker three basic functions may be distinguished:
- keeping the module operating in its maximum power point, independently of irradiance, module temperature and battery voltage, i.e. current-source nature,
- preventing overcharge of the battery
- preventing reverse current flow through the module during darkness.

Principle of operation

The basic assumption for a succesful application of this MPPT (as this MPPT consists of an up-converter) is the supposition that the maximum power point voltage V_m is always lower than the battery voltage V_b.

Since neither V_m nor V_b are constant, the MPPT should continuously bridge the gap between the voltage of the module in its maximum power point and the actual battery voltage. As a matter of fact, the MPPT generates an extra voltage on top of V_m, so that the output voltage of the MPPT equals the battery voltage. A necessary condition to charge the battery.

This upconversion of voltage is performed by the circuitry as depicted in figure 8.

Max. power point tracker.

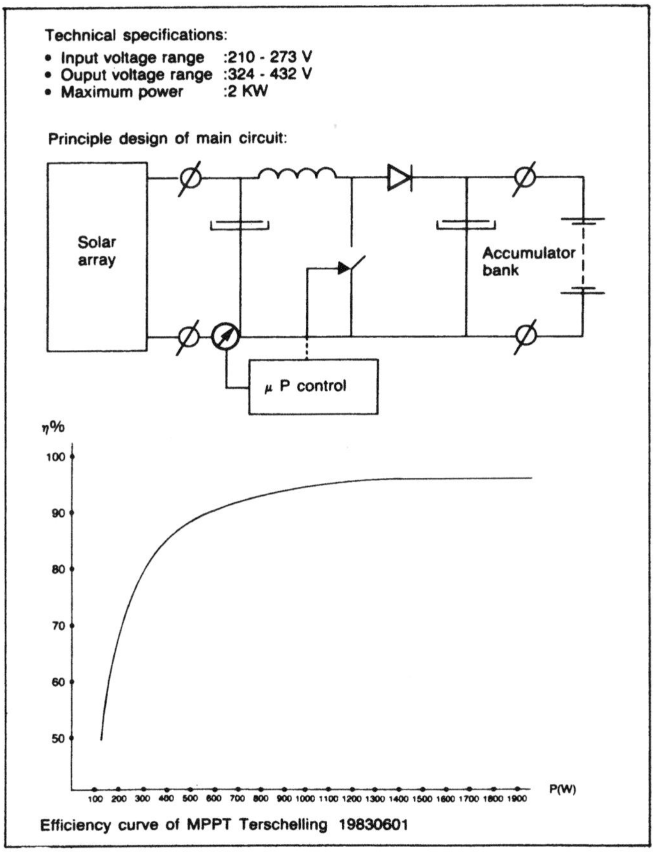

Fig. 8

The inductive coil will introduce an extra voltage on top of V_m due to Lenz-Law:
When the switch is closed, the current through the coil will increase. At

the moment the switch is opened, however, the coil will try to maintain the same current.

In doing so, it will induce a voltage V_c, which depends on the duty cycle of the switch, being the time, the switch is closed (t_{on}) in relation to the total cycle time (t_{on} + t_{off}).

Mathematically it can be proved, that the extra voltage V_c induced by the coil will be:

$$V_c = \frac{t_{on}}{t_{off}} V_m$$

where V_m is the actual voltage of the module.

Reversed current flow will be prevented by the diode in the circuit.

Protection against overcharge is accomplished by opening the switch permanently upon reaching the gassing voltage of the battery.

The switch is fast safe to the off state and external forclug to this state is possible.

The microprocessor checks whether the actual working point of the connected subarray is delivering max. output. Each quarter of an hour this optimization-procedure will be initiated by the micro-processor.

The technical specifications and efficiency as a function of the MPPT are mentioned in attached viewgraph 8.

3.2.2 Inverter

Working principle

As shown in fig. 9, a well-proven design is proposed. The main items of the system are a 60 kVA load-commutated current-source inverter and a 50 kVA synchronous motor. The latter provides the EMF, needed for natural commutation of the inverter in the stand-alone mode. One of the major features of this lay-out is its ability to support the surge currents which are to be expected in conventional electrical distribution systems. The inverter operates in the minimum extinction-angle mode in order to optimize its power factor. Thus a DC-chopper is needed for adaption of the battery-voltage. The chopper also provides additional protection in case of commutation failures.

A 40 kVA harmonics filter, consisting of three tuned branches, ensures low voltage distortion. At the same time it compensates lagging power factor of the inverter. In this way the motor rating, and so its power loss, can be kept to a minimum.

The system has two control inputs. The chopper duty cycle controls the transfer of active power to either the consumer or the mains. In stand-alone mode the frequency is a measure for the proper active power balance. The motor excitation determines the reactive power balance with the system. If the motor is kept running in the mains mode, it may be used for power factor correction. In this mode the system offers no-break capability at the expense of decreased efficiency.

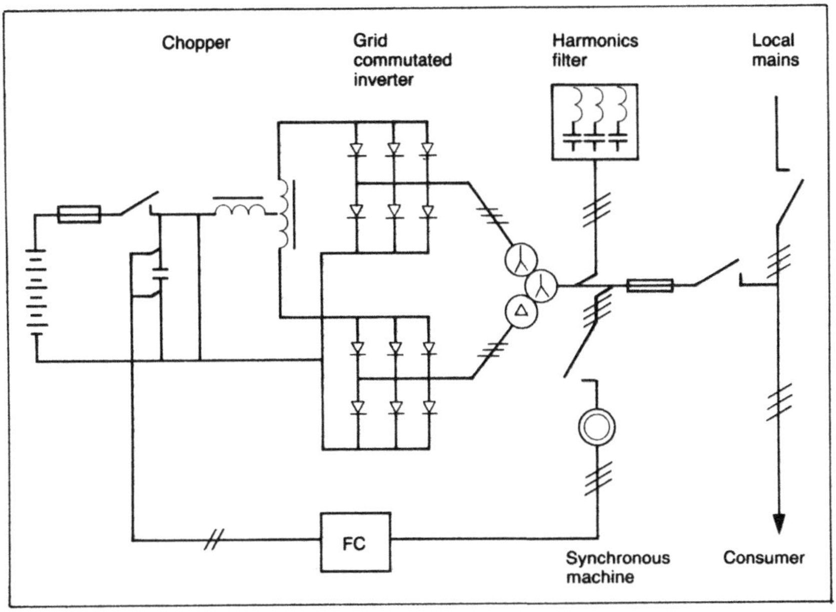

Fig. 9. Principle diagram of the inverter system

Main data

Input voltage: 324 ... 432 V DC
Output voltage: 380/220 V - 50 Hz
Max. Output: 50 kW active power and 35 kVA reactive power (with synchr. mach. in parallel)

Efficiency (including output transformer and choke)

Grid commutated without synchronous machine:
10% load 87 %
20% load 91,8 %
50% load 94 %
80% load 94,4 %
100% load 94,2 %

Stand-alone with synchronous machine:
10% load 65 %
20% load 75 %
50% load 87 %
80% load 89,4 %
100% load 89,8 %

As a feature the system can be used as an Uninterruptable Power Supply if the synchronous machine is stand-by in the grid connected mode.

3.3 Energy Management System (EMS)

Main purpose of the energy management system is the control of all energy flows and the collection of system-data.

The external flows are defined as flows exceeding the system-boundaries:

. the flow from the solar generator to the system
. the flow from the wind generator to the system
. the flow from the system to the public grid
. the flow from the system to the consumer grid.

The internal flows are the energy flows between the several system components. These components are:

. the maximum power point trackers
. the charger unit
. the two battery units
. the inverter

For interconnection of these components several internal grids are available:

. the AC-grid
. the DC-grid
. the reconditioning-busbar

If necessary the components are connected or disconnected to the grids by means of switches.

Lay-out of the control system

Basically two levels may be distinguished in the control system.

- the equipment level
- the operations level.

On equipment level the control of individual system components is performed. Therefore every component has "its own" dedicated controller. These controllers are completely autonomous and are suitable for stand alone operation.

So one might say, that on the lowest level all control functions are distributed over the components on which they have to be performed. The operations level has a coordinative function. It gives directives to the individual controllers in such a way, that an optimal total system performance is achieved.

As an example one might think of a setpoint which is calculated on operation level and given to a specific controller, which is responsible for keeping the physical value on the wanted setpoint.

In this way we get a hierarchical system, where the operations level gives directives to the equipment level, which will obey these directives.

The advantages of such a hierarchical distributed control system are twofold.

- first of all the system will get a modular structure. In future it will be very easy to expand the system without changing the controllers already available
- secondly the system will get a higher availability, because a break down of the vulnerable operations devices will not cause a total shut down, due to the autonomy of the devices on equipment level (a higher MTBF).

Due to modularity of the system an eventual error can be traced very rapidly resulting in a low MTTR. Apart from the control task, the operations level has also a monitoring function to collect all relevant data.

The data-monitoring has been integrated in the same microprocessor system as the EMS. The relevant data to be stored are fully in accordance with CEC guidelines and processed by a Penny and Giles tape-recorder type 7100. The P/G taperecorder is built together with the Satt microprocessor.

3.4 Array lay-out

The electrical lay-out of each array has been defined taking into consideration:

- dc-voltage-level (minimizing losses)
- design possibilities MPPT concerning up-conversion factor and power range
- service-ability
- module-size

For the 29 arrays equipped with MPPT this leads to 90 modules divided into 3 strings of each 30 modules (fig. 10).

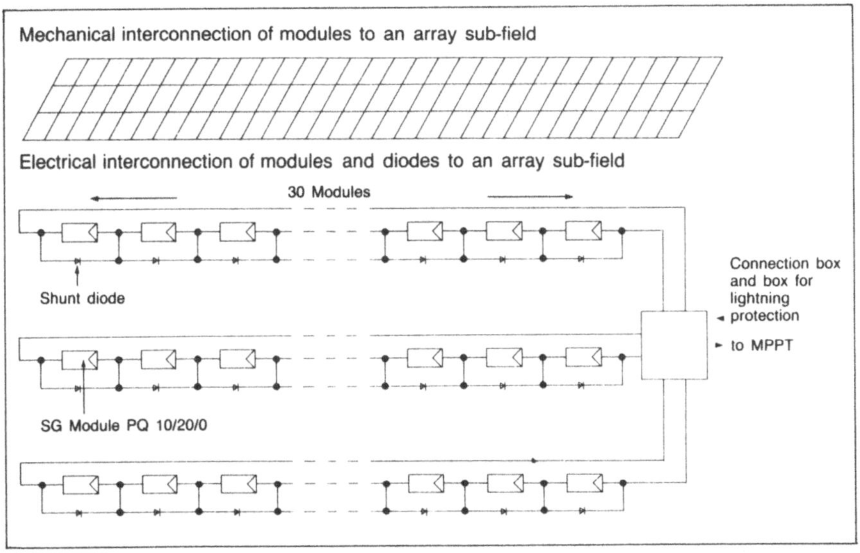

fig 10. Array sub-field (1,73 kWp) 30 Modules in a row 3 Rows parallel

In order to adapt the 30th array on battery-voltage, 3 strings of 46 modules are connected in parallel.

The calculation of the mechanical structures (fig. 11) is made on the basis of the Dutch National Code and the Monegon report.

In order to minimize structure costs, both reduction of windload on array height and the influence of a fence were taken into account, finally, a safety factor of 1,25 was applied.

The chosen structure is composed from pre-fabricated hotdip galvanised cold-rolled iron profiles. The design is especially effected to simple erection and a low weight in order to reduce erection and transport costs.

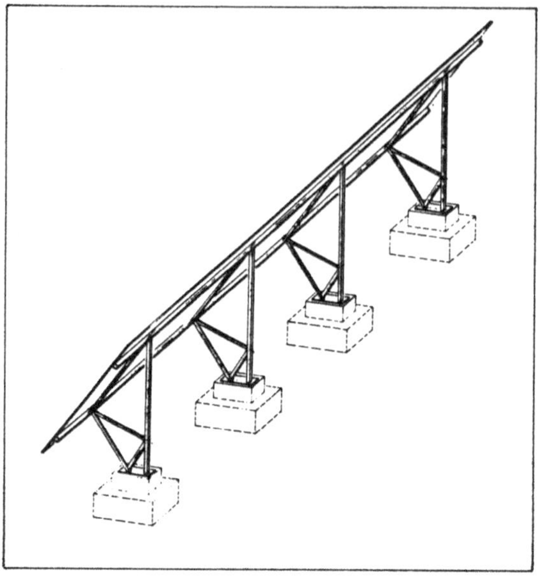

Fig. 11

Modules are manufactured with short leads and special connecting plugs for an easy and safe connecting to a string. The leads are laying in vinyl wire canals.

Earthing of the whole arrayfield is done by a grid of copper wire around the arrays with number of earthing rods: each array is connected at two sides to this grid.

3.5 Lightning protection

Due to the fact a direct stroke on an array can be expected once in approximately 200 years, the solar generator is only protected against indirect lightning.

This means all metal parts including the module-frames are directly connected to an earthing grid installed around the array field.

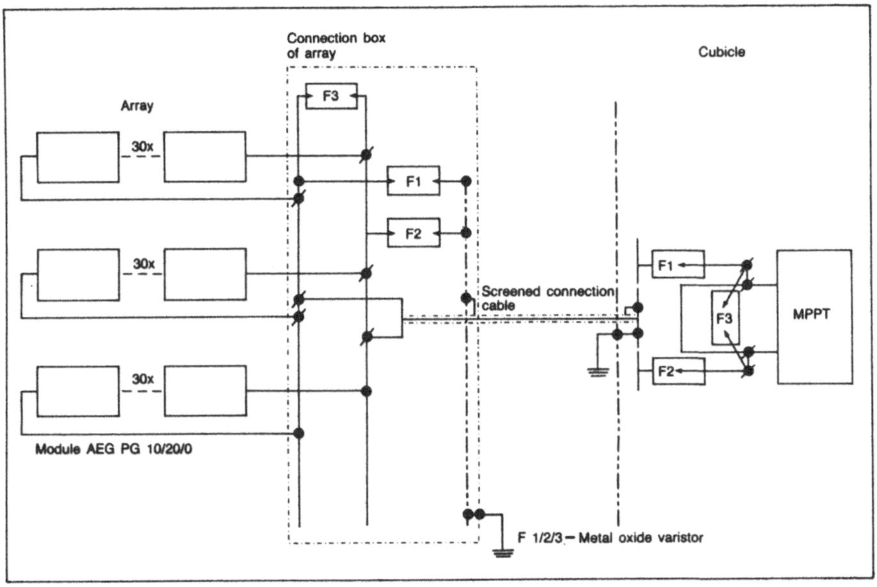

Fig. 12 Principle of lightning protection.

According fig. 12 metal oxide surge varistors are mounted in the connection box of each array and in the MPPT cubicle in the battery and control room (fig. 4)

The screen of the dc-cables connecting each array with the MPPT cubicle is connected to the earthing grid.

By means of these precautions the residual voltage will be limited to 1000 V assuming a direct lightning stroke at a distance of 100 m.

4. Experiences during Erection/Testing

4.1 Solar generator

During erection we found no major discrepancy with the initial design criteria:

- easy implementation of mechanical structures
- fast and easy mounting of modules
- first experiences during heavy storms proved mechanical feasibility (not too heavy but strong enough).

Some considerations for future design:

- tears on the surface of C-profiles occurred during the hot-dip galvanizing process can damage modules during fastening them.
- due to their extreme length (14,5 m) p.v.c. wiring canals expand even at moderate temperatures with the effect that closing lids jumped off. Additional fixation was necessary.

4.2 General remarks

The process to become all official permits and legalisation documents took much longer than expected.

Higher dc-voltages ($>$ 200 Vdc) are relatively dangerous. One should not underestimate phenomenons during switch and measuring procedures.

During recent testing load peaks of 20-30 kW occurred. Analyses showed this power-demand was initiated by the control-mode of certain kitchen-apparatus. In the stand-alone mode this would cause problems with the inverter. This means extra attention must be paid to the load-pattern especially for systems in the stand-alone mode.

Although frequent cleaning (once a forthnight cleaning of modules was supposed to be necessary, but experiences since November, 1982, showed that natural cleaning by rain, snow, etc., is more than sufficient) additional cleaning of modules increases output with max. 1%.

CONCLUSION

Apart from the consideration formulated by CEC to intall p.v. in order to prove in the community the possibility of fuel saving by renewables under actual circumstances, the next 5 years will show the technical and economical feasiblity of this system for applications in remote areas in the sun-belt countries.

The dual-purpose inverter system is designed to test the plant both in the stand-alone and the grid-connected mode.

An extensive additional test programme is in preparation.

AGHIA ROUMELI ELECTRICITY SUPPLY TO AN ISOLATED VILLAGE

Location, country	: Village of Aghia Roumeli Crete Island - Greece		
Rated peak power of array	: 50 kW		
Contract number	: ESC - P - 071 - F (G)		
Duration of contract	: December 81 - June 83		
Total budget of project	: 1,100,000 ECU		
CEC Contribution	: 350,000 ECU		(31.8 %)
	+ 70,000 ECU	(bonus)	(6.4 %)
Other funding bodies			
A.F.M.E.	: 285,000 ECU		(25.9 %)
Greek venture (PPC + ETBA + YEET)	: 285,000 ECU		(25.9 %)
SERI Renault	: 98,500 ECU		(9.0 %)
VARTA A.G	: 11,500 ECU		(1.0 %)
Contractor	: European Consortium Members SERI Renault Ingenierie (F) Public Power Corporation (GR) VARTAB Batterie A.G. (FRG)		

Project Leader : Bernard Aubert
c/o SERI Renault Ingenierie
2 av. du Vieil Etang, BP 19
F 78391 Bois d'Arcy
Telex : 695426 F
Telephone : (33) 3 043 81 34

Abstract

The aim of this project was to demonstrate photovoltaic generators as technically suitable, economically achievable and socially well-accepted solutions to meet today the energy requirements of remote villages out of reach of utility grids. In September 1982, an autonomous 50 kW photovoltaic generator has been installed for all electrical needs of Aghia Roumeli, a small village of 105 inhabitants located on the South West seashore of the Crete Island. Technological and system development has been devoted to each subsystem = array supporting structures, wiring features, power conditioning equipment, inverter and storage battery - Reliability, cost/efficiency ratio and automaticity were the key design criteria. As a result, this innovative photovoltaic power generator can be viewed as a prototype of future solar gensets that could replace conventional Diesel engine for stand-alone electricity supply of isolated communities : The Greek national utility "The Public Power Corporation" owns and operates the generator and its distribution network for their expected lifetime (20 years). The installation work was completed on October 15, 1982. Until now (July 83), the generator has been running quite satisfactorily. In June 83, mean AC production exceeded 200 KWh a day.

1. INTRODUCTION

The aim of this project was to demonstrate Photovoltaic Generators as a technically suitable, economically achievable, socially well accepted solution to meet the energy requirements of remote villages, out of reach of large utility grids.

The traditional way to cope with this problem consisted in on-site power generation by means of Diesel generating sets with high operating costs (fuel and maintenance due to remoteness). The solar alternative had to be investigated with an actual village in real operating conditions.

By the end of 1981, an autonomous 50 kW photovoltaic generator was decided on for the satisfaction of the electrical needs of AGHIA ROUMELI. It is a small isolated village (105 permanent inhabitants) located on the south western seashore of the Island of Crete (Greece). Located on a hill-side, Aghia Roumeli overlooks the Mediterranean Sea Southwards. It is surrounded on the other sides by the Samaria Forest, a protected area of an exceptional beauty.

This place is very far south in Europe (latitude = 35° 14' N, longitude 23° 58' E) and its mean daily global irradiation is estimated to be 4.8 kWh. However, due to its coastal situation, Aghia Roumeli has a quite tempered climate : the temperature varies in the range 25°C to 35°C during the summer.

Although the village is quite remote, this privileged location raised a natural inclination for tourism (visitors are evaluated to top 170.000 a year).

Aghia Roumeli is doubly isolated :

Geographically :

Aghia Roumeli can be reached either from the sea (by a daily shuttle boat during the tourist season), or by a footpath through the forest and the canyon of Samaria (5 hour walk).

Energetically :

The "Public Power Corporation" (P.P.C.), the Greek National Utility is not allowed to lay a line through the protected Samaria Park. Likewise, a coastal connection with the closest village is not envisaged, since it is necessary to lay a cable for more than 15 km under the sea ; this is not economically justifiable, if one takes into account the low level of the annual demand load of the village (less than 50 000 kWh a year).

The current supplied by the photovoltaic generator is used for the electrification of the homes, the shops, and the hotels for lighting, water pumping, cold stores and refrigerators and other usual domestic appliances.

Dedicated to the cause of the E.C. R and D programmes and to fulfill the aim of this project, each component is selected or developed according to its :

- energetic efficiency to increase the performance/cost ratio of the whole plant,
- reliability of power supply to obtain, as far as it is possible, the "guarantee of the kWh" and to minimize the maintenance costs,
- self-acting operation to allow the full automaticity of the photovoltaic generator (no attendance needed).

Moreover, our attention has been drawn to the need for the reduction in the "non-cell costs" which include the manufacturing, the transportation and the installation on site.

2. GENERAL TECHNICAL DESCRIPTION

System description is presented in the table below. Power management is essentially a control of the battery state-of-charge. The key role played by this parameter justifies a precise measurement of its value at anytime. This is ensured by the Varta-Logistronic control device.

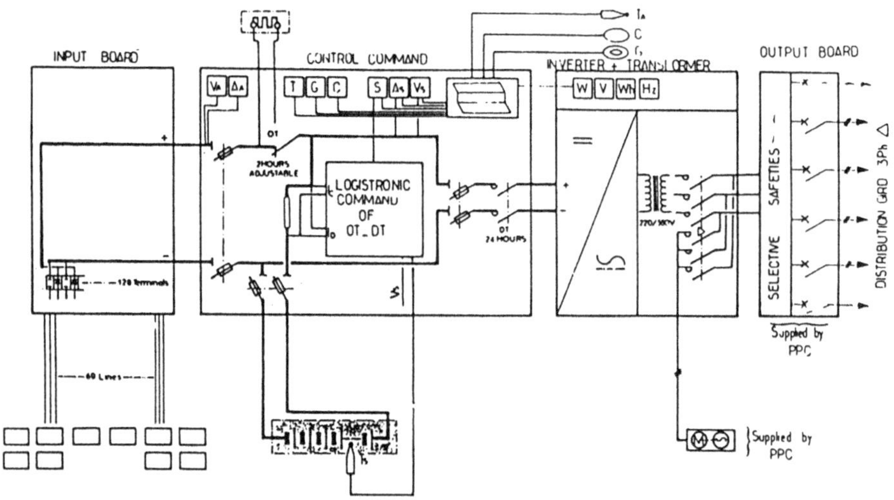

AGHIA ROUMELI _ A 50kW PHOTOVOLTAIC GENERATOR
COMMISSION OF THE EUROPEAN COMMUNITIES

The main specifications of the system are listed in the table below.

P.V. ARRAY	Peak photovoltaic power 50 kW (installed, at 1 kW/m^2 insolation and 25°C Cell's temperature) Mean daily A.C. load supply 70 kWh in December 180 kWh in July Composition : - 60 panels composed of twelve modules FPG 72 from France-Photon Tilt : 25°/horizontal
STORAGE BATTERY	Type : Lead-acid VARTA bloc Number of elements in series : 150 Capacity : 1500 Ah at C/10 hours Autonomy (at full load : 180 kWh/day) : - 2 days with an open-circuit array - 4 days under overcast sky conditions Voltage range : 292 V-353 V New design : - maintenance free for 3 years - low self-discharge : 3 %/month
INVERTER	Special "Solar" design for high efficiency and reliability. Rated permanent power : 40 kVA Type : Pulse Width Modulation transistored technology Output of plant : 3 x 380 V ac with step-up transformer with neutral Frequency : 50 $\pm$ 0,001 Hz Efficiency at COS Ø = 1 of inverter + transformer 93 $\pm$ 2 % at full load (40 kVA) 85 $\mp$ 2 % at a tenth of the full load No load consumption : - 200 W at the inverter output - 700 W at the step-up transformer output Conditions of use : - permanent operation (24 h/day) - temperature range 0-45°C - tropicalization : hygrometry 0 + 90 % marine atmosphere
OTHER FEATURES	- Fully automatic operation (no attendance needed) - Voltage regulation - Measurement of the battery state of charge by a new electronic device - System fully instrumented to allow all energy balances Solar irradiation, DC energy generated, storage contribution, AC energy supplied...

3. LINK TO APPLICATION : THE SITE AGHIA ROUMELI

The fair adequation between expected demand and solar source is to be noticed.

Solar source at Aghia Roumeli :

The mean daily global irradiation on a plane with a 30°/longitudinal tilt towards South was estimated to be $G_y = 7$ kWh.m^{-2} in July, $G_d = 2.5$ kWh.m^{-2} in December.

Demand profile of the village :

The brief table below gives an evaluation of the former daily consumption ensured by privately owned diesel gensets.

USERS	WINTER	SUMMER
34 homes	17	34
7 hotels	35	105
8 shops	12	40
TOTAL kWh	64	179

The major part of energy is used for lighting and cold producing appliances. One has to notice the fairly good adequation between the seasonal distribution of electric needs and the seasonal distribution of the solar global irradiation.

Dimensioning the plant :

It deals with the power vs time matching of the source with the demand. A computer analysis has been carried out to simulate the operation of the photovoltaic generator on an hourly basis and on a daily basis (See graphs of simulation on next page).

This computer simulation predicts upto what extent the solar generator can fulfil any scenario of demand as regards the insolation estimates. These predictions of performance were found to be in fair agreement with the actual performance measurements recorded so far (June,83) at AGHIA ROUMELI (See plotting on last page of paper).

AGHIA ROUMELI - COMPUTER SIMULATION

A 50 KW PHOTOVOLTAIC GENERATOR (MARCH 81)

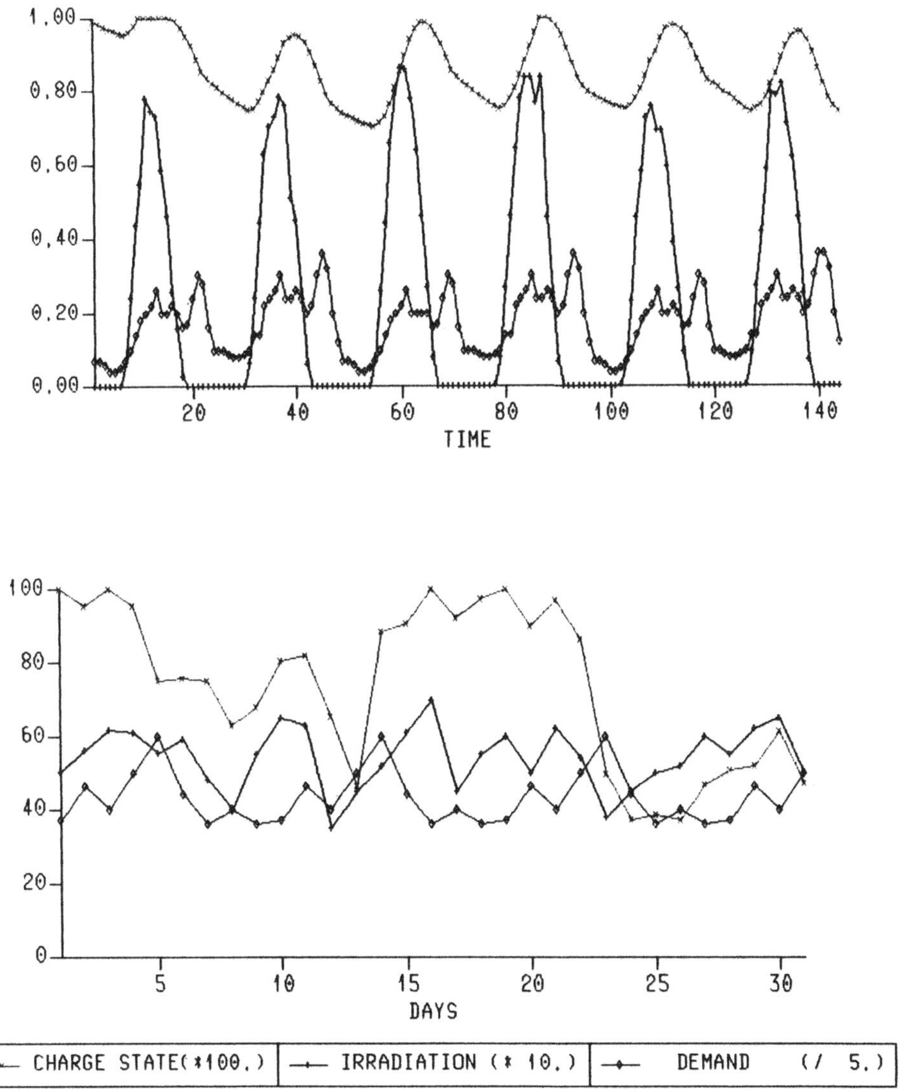

The main parameter - the state-of-charge of the battery was thus plotted to check out the proper dimensioning of the plant with respect to the input - the solar source and the output - the demand profile.

4. ARCHITECTURAL AND ENVIRONMENTAL ASPECTS

The general lay-out for the photovoltaic plant was drawn by a Greek architect after a visit to the site and according to the advice of SERI Renault. The main criteria for this lay-out was the environmental concern, the protection against vandalism and a possible flooding of the site, on the river delta.

The building and the surrounding fence are made by using the pebbles found on site ground, thus ensuring a perfect visual integration of the environment (see photography below).

5. OVERVIEW OF NEW TECHNOLOGICAL DEVELOPMENTS

In order to reduce the so-called no-cells costs of future medium size pv generators, as well as to enhance the systems' reliability, a considerable amount of technological and system development has been devoted to each subsystem. For instance the array supporting structures, the intra panel wiring and array wiring systems, the battery state-of-charge management, the power conditioning equipment including the inverter and its protection features.

Apart from this specific R and D work on components, it was the first time that such pv modules were used to form 350 V dc sub-array strings, also these new type of battery elements from Varta were never been used before on pv plants.

6. PHOTOVOLTAIC MODULES

The photovoltaic array is a field of 60 individual panels (or sub-arrays). A panel consists of a supporting structure and 12 modules connected in series. Each panel is linked to the electrical building via the "input paralleling board" which connects all of them in parallel.

The modules are manufactured by France-Photon. The module was developed specially for the pv pilot projects of the E.C. R and D programme.

Consisting of 68 round Ø 100 mm monocrystalline silicon wafers, the module is particularly adapted for large pv power plants. For instance, every 17 cell string is protected by a parallel diode included in the back of the module.

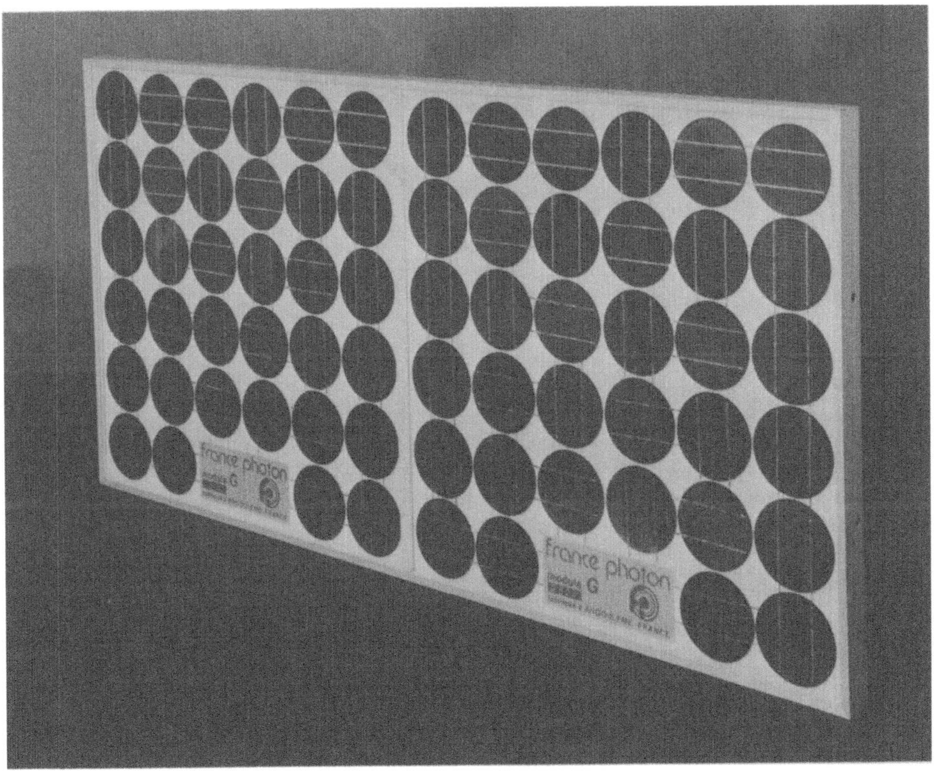

7. ARRAY SUPPORTING STRUCTURES

The purpose of the development work performed was the minimization of the total cost including manufacture (ex-works costs) subdivided into materials + labour + anticorrosion treatment + screwery and ground preparation, foundations, transportation, mounting on site, cabling. The optimized structure is presented below.

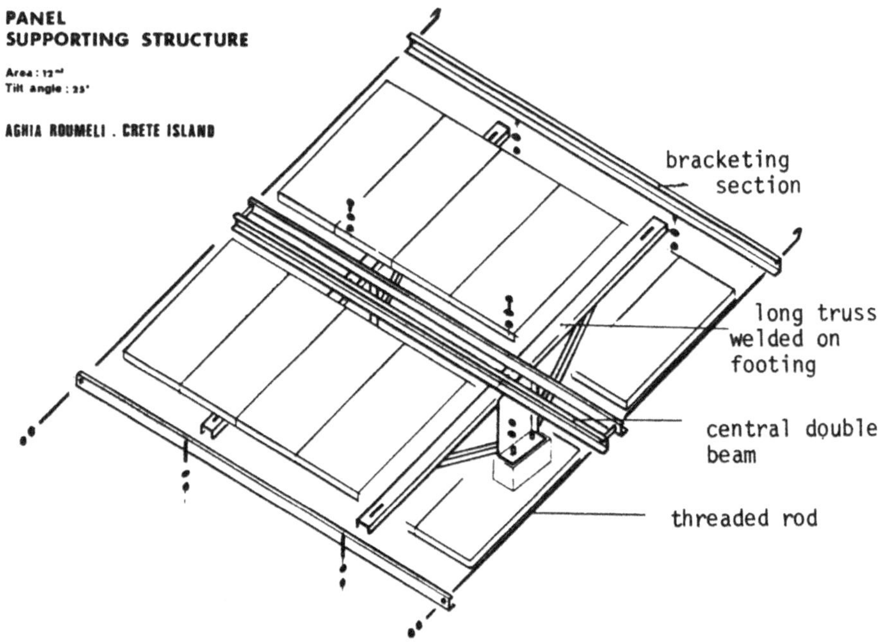

8. LIGHTNING PROTECTION

The lightning protection that was implemented for the pv generator at Aghia Roumeli was decided on after having estimated the probability of direct lightning strokes on the plant.

Probability of direct lightning strokes :

The method followed is described in the CESI study report TS 200-2-Milan, September 1981 prepared for the CEC.

In Crete Island the keraunic level (T.D.) averages 30 days a year. Since the Aghia Roumeli village lies on the seashore, one can expect that this average value exceeds the "spot" value for the site.

The earth flash density (N.g) is drawn from :

N.g = 0.023 T.D. = 1.9 flash/km^2/year.

The equivalent exposition area of the plant is : Aeq = 3800 m².

So, the probability of direct lightning strokes to the station is :

$$Ns = Ng\ Aeq = 1.9 \times 3800.10^{-6}$$

$$Ns = 0.0072 \text{ flash/year.}$$

And so, the mean time between flashes is :

$$Ts = \frac{1}{Ns} = 138 \text{ years.}$$

As a first result, it can be noticed that this figure greatly exceeds the expected lifetime of the plant Tp = 20 years.

Taking into account the comparison between the two figures, it can be assumed that the generator is very unlikely to be subjected to dangerous direct lightning effects. Consequently it was decided that a careful grounding of all support structures of the array through a 2 x 30 mm² cross section copper wire that nets 10 m by 15 m the entire array field, would be "suitable and sufficient".

9. STORAGE BATTERY

The energy storage of the photovoltaic plant is ensured by a stationary battery of a special industrial design : the VARTA BLOC Vb 2415 type. The battery consists of 150 elements connected in series. The elements are installed on floor stand in a free convective vented room.

Dimensioning of the nominal storage capacity was performed by using the dynamic simulation programme afore-mentioned. The inferred autonomy corresponds roughly to 2-4 days at full load.

Nominal capacity (rated in 10 hours) is 1500 Ah, nominal voltage is 300 V. The depth of discharge will be limited to 70 % of the total capacity. However, the "solar" operating conditions increase in practice the available capacity by 30 %.

Available capacity = Nominal capacity x 0.70 x 1.30 = 410 kWh at 20°C. See below the capacity curve, also discharge curve.

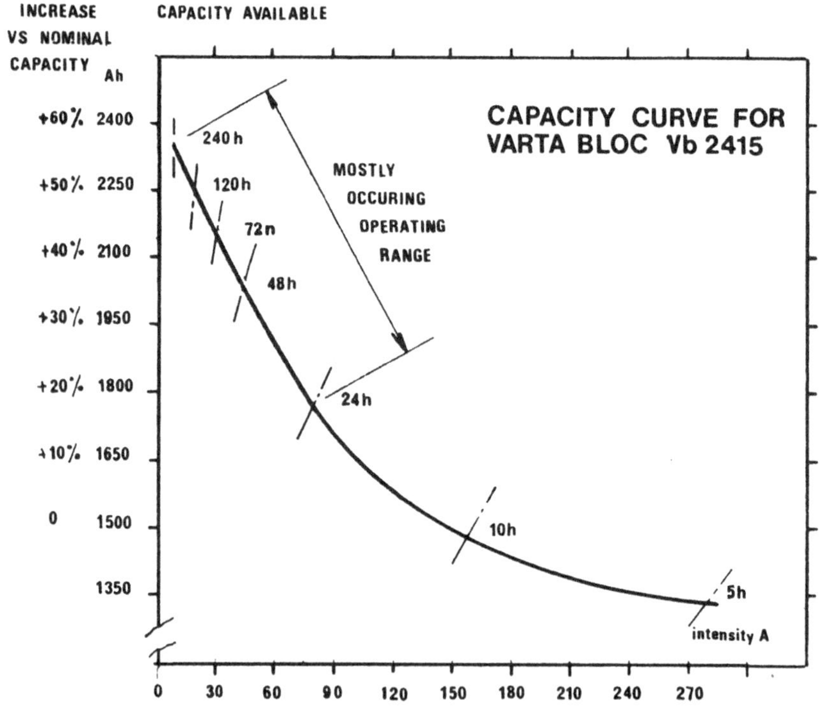

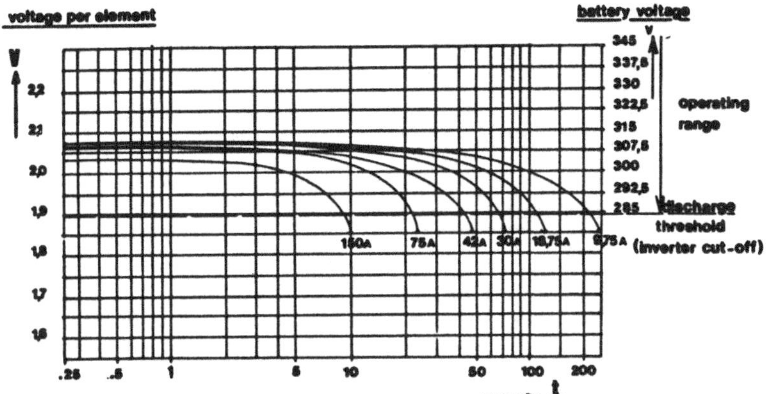

Discharge-graph for VARTA bloc-cells in dependence on the discharge-current → t/h

10. POWER CONDITIONING

The inverter is a three phased, specially developed device using new power transistors + the phase width modulation technology. Its main specifications are given in the table of chapter 2.

The use of newly developed transistors makes the switching possible over a wide range of voltage. The fixed modulation of the width with the monitoring or the current in the load, in addition to the cancellation of the usually used inputs regulating chopper, increases efficiency in particular at lower loads.

11. MONITORING SYSTEM

The monitoring system that was originally foreseen with the equipment to measure the following operating data :

IG : global irradiation measured and collected by a Kipp and Zonen pyranometer and integrator (every hour and every day print-out) plus a paper chart analog plotting.

IA : Irradiation in the array plane (25°/horizontal) integrated by a Haenni module and plotted on the chart recorded.

Ta : Ambient temperature plotted on the chart recorder.

Vb : Battery voltage displayed and plotted.

Ia : Intensity of the pv array displayed and plotted.

Ib : Intensity of the lead-acid battery displayed and plotted.

kW : Output power of the plant displayed for each phase and plotted.

Ahin) the Ampere-hours "in" and "out" the storage battery displayed by
Ahout) the logistronic device.

kWh : The energy produced each day displayed.

H : Number of operating hours.

This basic data is very important to follow the plant operation and to help to identify any possible occurence of defect.

12. SYSTEM INTEGRATION AND INSTALLATION

Civil works (building, array foundations) were performed by local subcontractors by August 82 and was terminated by October. Plant installation was scheduled during August 82. It lasted 4 weeks, by hiring four mechanics and two electricians from PPC. Installation has been performed according to the recommendations, advice and under the supervision of the designer SERI Renault. Mounting time of all operations has been recorded in order to, on the one hand, measure the progress made on this plant relative to standard design, and on the other hand, to collect data concerning the installation costs of such medium size photovoltaic plants.

The operation of the plant during its expected lifetime (20 years) will be handled by PPC. In the first years of operation, SERI Renault and PPC will collect jointly the operating data and findings.

Data monitoring equipment is presented in the system structure chapter 2. For such an isolated site, a paper chart recorder was selected for its reliability, ease of operation and "visual character".

13. CONCLUSIONS AND FIRST RESULTS

Due to the availability of electricity 24 hours a day, this project was expected to bring an important amount of experience and knowledge on technical, economical and social aspects. For instance : how will it increase the demand profile of the village ? What will be the price of the kWh produced ? What is the reliability, relative to conventional Diesel gensets ? In other words, what is the future of these intermediate photovoltaic power plants ?

After 9 months of operation (June,83),very positive answers can already be given to all of these questions : the maximum mean demand of the village was at 220 kWh a day in June 83 without causing the battery storage to run-dry. And so the first rough estimates of the cost of the kWh produced average 2 ECUS over an expected 20 year lifetime.

Most of all, the reliability has proven to be excellent so far (9 months). As a matter of fact no significant breakdown was encountered although more than 15 MWh has been supplied to the village grid.

CHART RECORDING OF THE INPUT/OUTPUT OF THE AGHIA ROUMELI PV POWER PLANT

This graph showing the power conversion versus time is in fair agreement with the predictions of the computer simulation (see plottings of chapter 3).

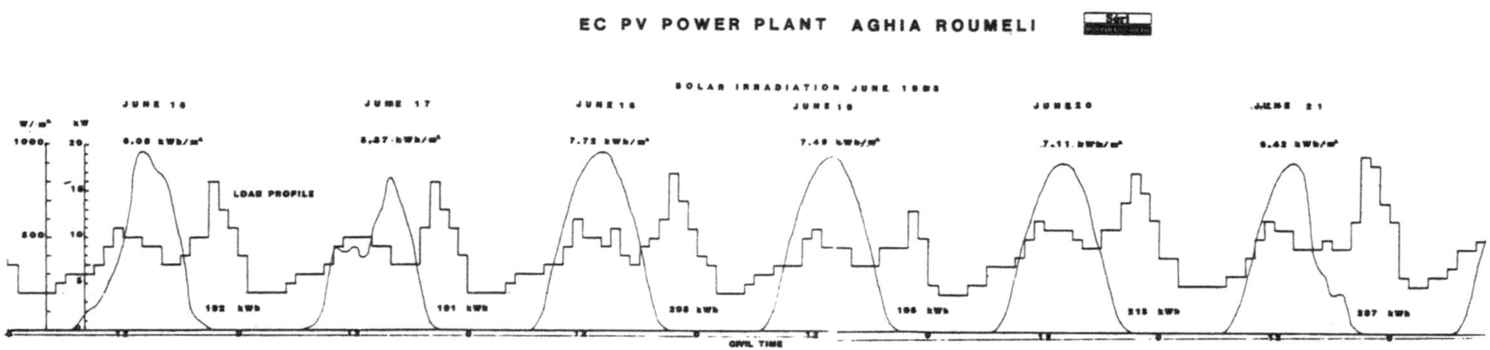

WATER DISINFECTION SYSTEM AND COLD STORE

Location, country : Giglio Island, Italy

Peak power of array : 45 kW

Contract number : ESC-P-065-I

Total budget : 1 620 650 ECU

Sponsors besides CEC : ENEA (Comitato Energia Nucleare ed Energie Alternative) - Regione Toscana - Commune Isola del Giglio

Members of association : - OTOMELARA SpA
- NUOVO PIGNONE SpA

Project Leader : Maurizio Tomassini
PRAGMA SpA
Via Po, 25/A - 00198 ROMA

Summary

The plant comprises two array subfields with a total rated power of 45 kWp.
Each array subfield is completely independent of the other and supplies the following loads :

a) water disinfection system, based on a commercial ozonizer (15 kWp);
b) a cold store for the conservation of foodstuffs (30 kWp).

This document outlines the design concepts which make for optimal utilization of solar energy and the basic configuration which it has been decided to adopt.
Particular attention has been paid to the control of critical process parameters in the water disinfection system. The cold store system, which has been designed by OTOMELARA to operate on the photovoltaic power available, has been optimized by using a DC moter compressor which has a combined control for the motor speed and cylinder loading, according to the insolation that is available. It has been decided to use eutectic plates to reduce the consumption of the auxiliaries and ensure a more uniform cell temperature.
PRAGMA SpA has responsibility for the project management.

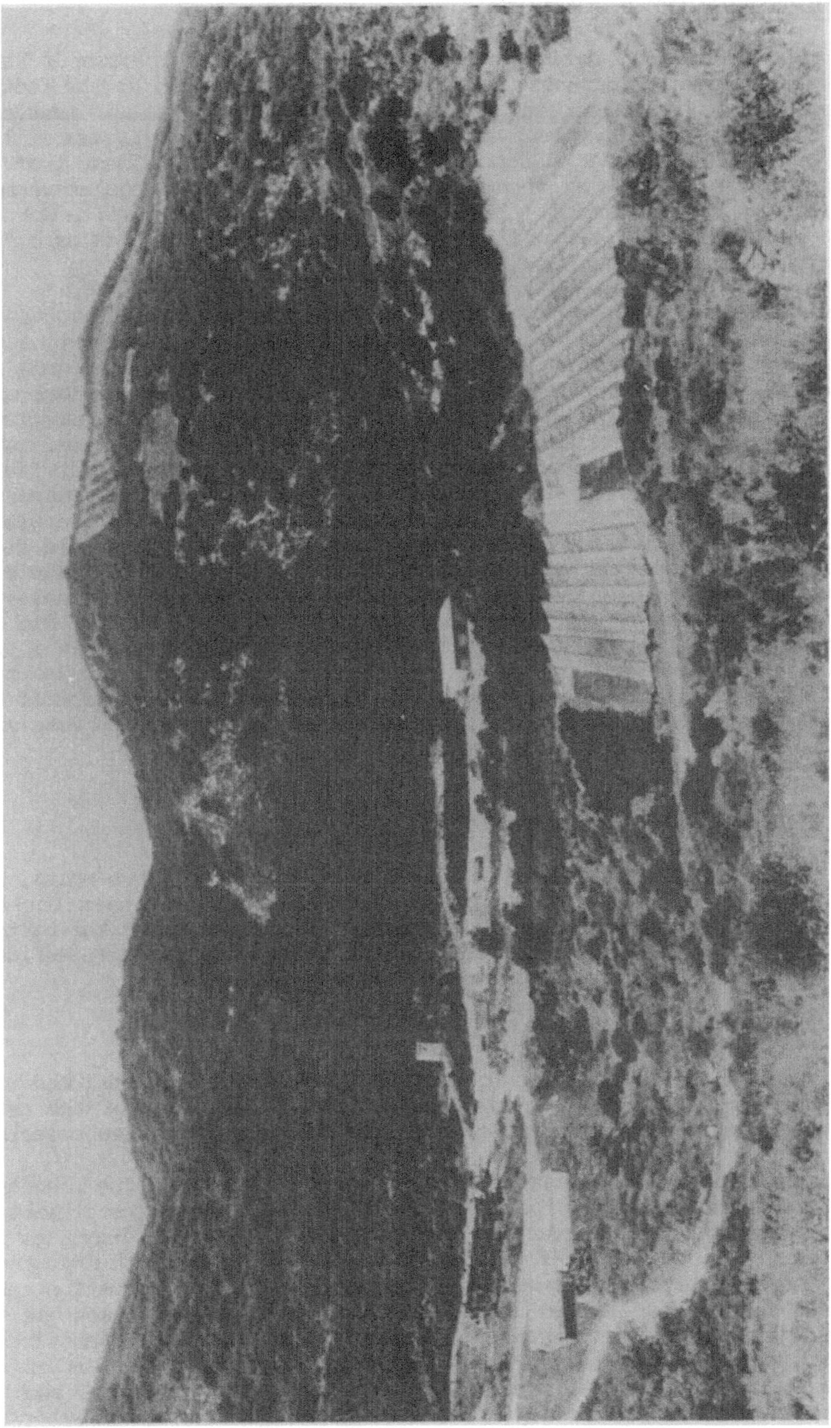

1. DESCRIPTION OF THE PROJECT

It has been decided to set up the photovoltaic plant in Fosso di Valle Ortona. The photovoltaic panels will be located on a hill, at the foot of which the Giglio Island local authority plans to set up an industrial estate. The cold storage plant and the water disinfection system will be set up in this area, at a distance of 100 m, as the crow flies, from the photovoltaic field. The gradient varies and there are irregular manmade terraces. It has accordingly been necessary to excavate and adapt the site to accommodate the supporting structures. Figure 1 contains a "Key Plan" of the plant.

The photovoltaic field, which comprises two subfields, supplies the two user systems separately. Fifteen of the 45 kWp are assigned to the ozonizer and its ancillary equipment. The low-voltage bus bars are connected to the charge control system, which is located in the same area as the ozonizer. The load comprises the ozonizer and the associated controls for the ozone released into the atmosphere. The control system and the data-recording instruments are supplied by the photovoltaic field, with mains back-up. There are a considerable number of measuring instruments. The remaining 30 kWp are transmitted to the cold storage plant. This comprises two distinct loads. The first consists of the cold store auxiliaries (condenser fan and the motor stator coil), plus the requisite inverter. The second is the cold store compressor motor, which has a DC power supply. A system of batteries is used to store energy for the auxiliaries. The control system functions by gearing the speed of the motor and the number of compressor cylinders in operation to the power available from the photovoltaic field. We give below a description of the basic configuration of the project, including adjustments which were made during the project development stage.

2. DESCRIPTION OF THE USER EQUIPMENT

It has already been mentioned that the plant is to supply two units, one for the conservation of perishable foodstuffs and the other for the purification of water for drinking purposes. Both units and the battery storage are to be located in the same prefabricated buildings (see Figure 2). We propose to describe the two user units separately.

2.1 Ozonizer

The decision to use an ozonizer to disinfect the water was mentioned in a previous report (1). We will now give a brief description of the principle underlying the functioning of the ozonizer that has been selected.

The ozone is produced by means of an electrical discharge. The discharge is not direct (voltaic arc), so as to avoid burning other constituents of the air. The operating voltage is around 15 kV. When the ozone is put into contact with the water (for a period of time - measured in minutes - which is governed by a whole series of parameters), rapid oxidation takes place, and bubbles are formed in the contact tower. This causes all the impurities present in the atmosphere, including bacilli, bacteria, viruses and spores, to be burnt up and, at the same time the molecules of any flavours or odours are destroyed, and inorganic constituents - such as iron, manganese and ammonia - are oxidized, thus sterilizing the water and making it fit for drinking.

The ozonizer has to function continuously for long enough (at least 10 hours) to ensure that the percentage of ozone produced is correct and consistent. The equipment has to operate at a constant throughput in order to avoid any infiltration due to lack of over-pressure. Inside the equipment there is to be an open-cycle cooling system (using water produced by the process since the coil is sufficiently clean). In addition to the ozonizer, there will be a system for monitoring the quantity of ozone dissolved in the water and another for monitoring the percentage of ozone released into enclosed spaces (it is toxic in high concentrations). Without going into details about the design of the ozonizer, it will suffice to say that the parameters which ensure the purity of the water include the air temperature (this influences the percentage of ozone released into the atmosphere), the temperature of the water (the optimum temperature being around 10 to 20°C), and the period of time for which they are accordingly in contact. By increasing the percentage of ozone dissolved in the water (and therefore the electrical output of the equipment), it is possible to reduce the length of time they are in contact, within the constraints of the maximum temperature threshold and the minimum time threshold.

Our ozonizing equipment comprises the following units :
- ozone production unit;
- mixing unit;
- gasification and degasification unit.

2.1.1 Ozone production unit

The following operating conditions are required for the system :

- temperature : 30°C max.
- relative humidity : 60%.

The main specifications are :

- ozone production : 15 g/h min.
- output of ozonized air : 0.8 Nm^3/h approx.
- ozonizer air pressure : 0.5 bar max.
- regulation of ozone rate : manual
- power consumption : 1.0 kW approx.
- cooling : by water
- power supply : 220V/50Hz/1 single phase
- consumption of cooling water : 0.024 m^3/h approx.

The main components of the generator are :
- the ozone generator
- the air-drying plant
- the electrical operating, control and alarm devices.

All the equipment is enclosed in a metal cabinet measuring :
- 1.85 m in length
- 1.80 m in height
- 0.50 m in depth.

2.1.2 Mixing unit

This is used to dissolve the ozone/air mixture in the water that is to be purified.

The principal components are :
- Centrifugal pump to create an overpressure :
  . capacity : 1.3 m^3/h approx.
  . head : 1.75 bar approx.
  . power : 0.37 kW approx.
  . motor speed : 2 800 rpm
  . mains power supply : 220 V/50Hz/single phase
- AISI 304 mixing tank :
  . diameter : 400 mm approx.
  . depth : 1 000 mm approx.
- Ejector with the following specifications :
  . suction capacity : 0.8 Nm^3/h approx.
  . drive water : 1.3 m^3/h approx.
  . mixing ratio : 0.68 approx.
  . maximum return pressure : 0.5 bar approx.
  . required intake pressure : 2.2 bar approx.
- Retaining valve fitted to the ozone pipe
- Pressure gauge with cock : 0:2 bar.

2.1.3 Gasification and degasification unit

This unit, comprising the following components, causes the ozone to react with the oxidizable substances.
- Steel reaction tank
  . diameter : 500 mm approx.
  . depth : 1 800 mm approx.
- Automatic breather pipe
- Ozone distribution device.

2.1.4 Power consumption of the equipment

By analysing the power used by the individual components of the system, it has been possible to make a realistic calculation of the consumption.

If account is taken of the consumption of the control system, the figures are as follows :

- ozonizer	: 1.5 kW
- transducers	: 0.2 kW
- control unit	: 0.2 kW
TOTAL	: 1.9 kW

2.2 Cold store

In reference document (1), we explained why it had been decided to adopt a refrigeration system based on eutectic plates. We now propose to give a detailed description of the plant.

The refrigeration plant is a Freon-22 cycle plant, with "cold" storage in eutectic plates, and it has been designed having regard to the following assumptions :

- outside temperature : + 35°C approx.
- energy supply available from the photovoltaic field : 20 kW for 3 hours per day
- reserve of "cold" : 2 days, by means of 120 eutectic plates (258 000 frigories)

Under these conditions, the regfrigeration compressor which is powered by a 20 kW electric motor, can supply on average approximately 40 000 F per hour or 120 000 F per day, during the "cold" storage stage.
During the "cold" storage stage, energy has to be supplied to the compressor (20 kW) and the electrical fan (2.4 kW) of the air condenser.

An exceptionnaly large condenser is to be used, in order to confine the use of the fan solely to days when the ambient temperature is exceptionally high.

The refrigeration compressor is of the R22 single-phase-cycle 4 cylinder type, with a power reduction device having the following specifications :

Frigories per hour	Temperature		rpm	Power consumption (kW)
	Expansion °C	Condensation °C		
46 000	- 5°C	+ 50°C	1 200	17.2
42 000	- 10°C	+ 40°C	1 200	14
21 000	- 10°C	+ 40°C	600	7

The part of the prefabricated building housing the cold store unit is illustrated in Fig. 3.
The store is insulated with polyurethane resin foam with the following properties :
- thickness : 120 mm
- density : 28-40 kg/m^3
- conductivity at 0°C : 0.018 kCal/m/h/°C

The resultant thermal conductivity coefficient is 0.16 kCal/m^2/g/h.

The operating conditions are accordingly as follows :
- temperature of regrigerating fluid : - 8°C
- temperature of plates : 0.5°C
- temperature of refrigerated store : 4°C

3. ARCHITECTURAL AND ENVIRONMENTAL ASPECTS

It has already been mentioned that the photovoltaic field is to be set up on the slopes of a hill in the Fosso di Valle dell'Ortana district. As this hill overlooks the area where the Giglio Island local authority plans to set up an industrial estate, it is the ideal site for a building using the photovoltaic plant. Although the area is to be an industrial estate, an attempt has been made to blend the structures in with the surrounding area. The area excavated to accomodate the structures supporting the panels slopes at the same angle as the average slope of the hill, so that the panels will not protrude above the Mediterranean scrub which grows on the hillside.

Immediately beneath it is the area where the cold storage and ozonizers systems and the battery room are to be located in one fairly small prefabricated building with a total surface area of 240 m^2 (including the auxiliaries room and the canopy over the ozonizer). There is to be a "screen" of vegetation, both to camouflage the outline of the prefabricated building and provide it with some shade.

It is also proposed to construct a road to link the plant with the inhabited part of the island; this will make for easy communications and therefore the functional use of the plant. There is also to be a road connecting the area occupied by the users with the photovoltaic field; the route which this road will take has been so selected as to interfere as little as possible with the landscape of the area.

4. NEW TECHNIQUES WHICH HAVE BEEN DEVISED

We will now describe the techniques which have been devised specifically for the plant. These are designed to obtain the maximum possible power from the photovoltaic field.

Attention has centred on the following aspects : maximum power tracking (MPT) and the development of a battery charging device (DC/DC converter) capable of ensuring this.

Maximum power tracking consists in gearing the operating voltage and current of the photovoltaic field to the maximum power which can be generated at a given moment. This can be achieved by varying the load impedance or the apparent impedance of the primary circuit of the DC/DC converter.

The converter is accordingly a variable ratio transformer in which the voltage ratio can be continuously varied.

However, since the voltage at the secondary circuit varies according to the state of the battery, the value that can be controlled is in this phase the input voltage (or current).

The control unit (UC) will generate a command which is proportionate to the input voltage required (this can be selected) and the DC/DC converter will immediately adjust the switching characteristics (from conduction to cut off, of the electronic switches in the converter, to provide the requisite voltage.

The standard modules located in the photovoltaic field emit the signals for short-circuit current and no-load voltage and also the temperature of the cells. When the signals are processed by the control unit, it is possible to determine the maximum power point from the insolation and temperature at the time the signals are recorded. Thus, in the case of the ozonizer, the control unit sends an analog signal to the DC/DC converter so that it can adapt its operation to the MPT, and in the case of the cold storage system, it sends signals to adjust the load to MPT conditions, by acting on the motor stator coil or bringing in or cutting out cylinders in the compressor.
The main operating criteria of the ozonizer and cold storage system will now be discussed separately.

4.1 Ozonizer system

The operation of the DC/DC converter in the MPT phase has already been described. The input voltage (from the photovoltaic field) must at all times be higher than that of the batteries. If the control unit requires a lower voltage than the batteries for MPT, the operating point of the field is so adjusted that it has a higher voltage than the battery at that particular moment. The DC/DC converter is accordingly of the BUCK type, that is to say it operates with higher input voltage than output voltage.

When the DC/DC converter is not functioning in the MPT phase, there are two operating conditions :

a) Minimum power threshold :
The minimum power threshold is that which is capable of supplying adequate power to compensate for transmission losses from the field to the users with a higher input voltage at the DC/DC converter than the minimum required for BUCK functioning (in the region of 143 V per battery in the charging phase).
If in this phase the voltage should prove to be below 143 V for MPT, the DC/DC converter adjusts the operating point to 143 V and does not follow the MPT logic until the power from the photovoltaic field is inadequate to ensure an operating point with a voltage above 143 V.

b) Maximum power threshold (battery charging phase)
When the battery is charged and the power supply from the photovoltaic field is higher than that needed to keep the battery charges plus the power required for the load (ozonizer plus data acquisition system plus auxiliaries), the DC/DC converter abandons the MPT logic and switches to the battery charge phase, shifting the operating point to a higher voltage and a lower current so as to match the power supply with the level required for the load plus battery charging.
When the DC/DC converter does not reply to commands for the MPT logic issued by the control unit, it sends to the control unit and the data acquisition system signals from the battery charging logic being applied.

The DC/DC converter used in the ozonizer system has the following electrical specification :

- Input voltage : 90 – 350 V, DC
- Output voltage : 90 – 160 V, DC
- Power rating : 15 kW
- Maximum input current : 100 A $\pm$ 10 A
- Minimum current : 2 A

4.2 Refrigeration system

The operating criteria for the DC/DC converter in this system and the MPT are the same as those described for the ozonizer. However, in this instance the configurations vary in accordance with the level of insolation and the power that is accordingly available from the field.

The DC/DC converter supplies only the control unit, the batteries and - via a chopper-amplifier - the motor stator coil.

The electrical specification of the DC/DC converter is as follows :

- rated voltage at peak power : 210 V, DC
- permissible variation in voltage : ± 20%
- maximum voltage obtainable from the photovoltaic field : 330 V, DC
- minimum current : 5 A
- maximum current : 60 A
- maximum power consumption : 15 kW
- efficiency : 95%

5. MODULES

The modules used in this plant comprise 72 cells connected 12 by 12 in series to form 6 strings in parallel.

There are other connections between the strings in the module (partial parallels) to minimize the consequences of clouding over or cell failure. Further, in order to obtain the maximum power available from the field, it has been decided to use modules with two separate terminals (positive and negative); hence the modules will have two terminal boxes located at the end of the short side of the panel.

This layout means that it is possible to shut off a string (comprising 48 modules) from the same side of the photovoltaic field, as shown in Fig.4, and furthermore to reduce the length of the connecting cables.

Fig. 5 illustrates the electrical specification of a standard module.

6. PHOTOVOLTAIC FIELD AND WIRING HARNESS

The photovoltaic field consists of 864 modules, plus 4 modules which are used to receive testing signals. Fig. 6 shows the layout of the strings. It is apparent that there are actually two photovoltaic fields, comprising 576 modules (cold store) and 288 modules (ozonizer) respectively.

To keep down mismatching and to avoid the need for complex procedures for selecting modules, it has been decided to construct the photovoltaic field from 18 strings in parallel, comprising 48 modules in series. Thanks to the minimal loss of the MPT voltage obtained on the production modules, it is possible to select modules solely on the basis of current.

As they comprise classes with a current range of 0.2 Ampères, in theory it is possible to keep mismatching losses extremely low.

The average current and voltage at the MPT of a string will accordingly be 240 V and 10.9 A, the average power therefore being 2 616 Wp.

The field supplying the refrigeration system will accordingly comprise 31 392 Wp while that supplying the ozonizer will be 15 696 Wp, assuming that AM = 1.5 and T = 25°C.

The layout of the wiring harness (Fig. 4) is designed to reduce to a minimum voltage drops in the connecting cables. The wiring harness between the modules is in fact confined to small sections of cable from one module box (+) to the next (−). From the terminals of each string, the power cables lead direct to the electrical panels on the field which contain the end of string diodes and the protective devices shown in Fig. 7 for the refrigeration system, and in Fig. 8 for the ozonizer.
Furthermore, a bypass diode located in parallel to three modules protects the modules against "hot spot".

7. STRUCTURES

A broad description of the type of structure is given in reference document (1). We will now describe the variants which were decided upon following attempts to construct and assemble a prototype. Fig. 9 shows typical dimensions and details of how the foundations are secured, connections between one section and another and between the module frame and the section.

The foundations are to be secured by means of expanding plugs which simplify assembly and afford a high degree of reliability. The module frame is connected direct to the relevant section by means of a purpose designed bolting system. All these techniques cut down assembly time, and accordingly reduce costs. The entire structure is made of Fe 37 hot-dip galvanized steel and the bolts are made of DIN A2 stainless steel.

8. LIGHTNING PROTECTION SYSTEMS

Both the area of the photovoltaic field and that of the prefabricated building are to be equipped with lightning protection systems. These will comprise protective cables suspended from cement posts at the shorter sides of the photovoltaic field. The best means of protecting the prefabricated building is a "Faraday cage". This comprises an earthed copper or galvanized steel lightning conductor system at the corners of the building.

9. POWER OFFTAKE AND BATTERIES

As it was decided to have separate power supplies for the ozonizer and the cold store, two separate systems had to be devised to process the energy produced by the photovoltaic fields.

9.1 Ozonizer system

Figure 10 contains a diagram showing how electricity is supplied to the ozonizer.
The photovoltaic field supplies a 240 V, 65.4 amp. (AM 1.5 at $5^{o}C$) DC current.

The electricity is led via the aforementioned battery charge control to the batteries and the ozonizer, the current being converted from DC to AC by means of an inverter and an output filter before it reaches them. An excessive wave distortion might cause vibration and overeating in the asynchronous motor used to blow the air. As it is proposed to use a conventional ozonizer, there is no need to adapt the plant to a supply power at a frequency other than 50 Hz (this would entail costly additional work).

The inverter is designed for a power supply of 3.5 kW with 95% efficiency. It is transistorized and supplies a 6-step wave which is subsequently filtered. The output frequency is 50 Hz $\pm$ 1 Hz, and the voltage may differ from the rated values by 10%.

Between the inverter and the ozonizer there is a static switch for a mains power supply. This part of the photovoltaic field can also be used to supply the control unit and the data acquisition system.

Given the various requirements which have to be met :
- maximum power tracking;
- the supply of the requisite auxiliaries;
- guarantee of a significant number of battery life cycles;
- two days' reserve power supply;

the battery specification had to be as follows :
Capacity per element : 1 600 Ah per discharge in 10 hours
Number of elements : 60
Rated voltage (60 elements) : 120 volts.

9.2 Refrigeration system

Figure 11 contains a diagram showing how electricity is supplied to the refrigeration system.
The photovoltaic field supplies 240 volts 130.8 A (AM 1.5 at $25^{o}C$) direct current.

A number of operating configurations are possible, depending on the level of insolation and therefore on the power supply available from the photovoltaic field. These are :
a) power supply to the battery charge control alone
b) power supply to the motor
c) power supply to the motor and inverter
d) power supply to the BCC, motor and inverter.

The battery charge control operates with input voltages of 210 V, DC $\pm$ 20%.

The battery charge control comprises transistors with an efficiency of over 90% and it controls the battery charging current, affording quasi-maximum-power-tracking. The battery is to be the lead-acid type, comprising 120 2V elements, with a maximum discharge of 40% (60% charge).

Figure 12 is a block diagram of the inverter, which is a 12 kVA model (it has to be able to withstand the additional start-up currents of the fan motors), with 95% efficiency in rated conditions. The output wave is 6 step (PWM), and a single bridge is used. The output voltage is 220 V three-phase, with a tolerance of 10%, and 50 $\pm$ 5 Hz. The winding is triangular. The input voltage may vary from 240 to 340 V.

The emergency and torque control block allows for protective action in the event of a net three-phase short circuit at the converter terminals and the adjustment of the electrical torque. There is accordingly considerable scope for varying the fan load by varying the output from the converter.

The back-up switch block is used to trigger te back-up mains supply. The motor stator coil is supplied by the same line. Photovoltaic power is supplied direct to the motor, when there is an adequate degree of insolation, and this load is sufficient for the cells to work at maximum emitted power, by means of a double loop :
- regulation of motor speed
- number of cylinders in operation.

The aim of regulating the motor stator coil is to increase the speed (Figure 13) so that it functions at maximum speed when the maximum power is supplied by the photovoltaic field. When the speed drops because the power supplied by the photovoltaic field has changed, one of the cylinders of the volumetric compressor will be cut out, thus reducing the load and restoring optimal conditions.

The 24 kW, 1 500 rpm-rated motor has a magnetic pick-up on the shaft, which "reads" the speed (covalent generator) and the 4-cylinder compressor. The eutectic plates are served by the compressor and refrigerate the cold store.

10. DATA MONITORING AND ACQUISITION

Data acquisition is carried out by VIDIMAC 8000 microprocessor, which receives signals from both the refrigeration system and the ozonizer. The VIDIMAC 8000 also controls the ozonizing unit, although an independent unit is provided for the control of the refrigeration system.

Wherever possible, data monitoring and acquisition sensors in the area of the photovoltaic field have been concentrated inside the busbars located near the photovoltaic panels (see Figs. 14 and 15).

We give hereafter the principles underlying data acquisition and monitoring for the two systems powered by the photovoltaic field.

10.1 Control of the refrigeration system

The purposes of control in this system are :

a) to produce analog signals to pilot the chopper-amplifier and the inverter; the current supplied to the fan can be geared to the energy available from the photovoltaic field at any given moment.

b) To produce digital signals :
   1 - to switch on or off the electronic or electro-mechanical switches used to modify the load by dividing the current from the photovoltaic field among the DC/DC converter, the motor, and the fan;
   2 - to vary the number of cylinders in operation in the compressor.
c) To ascertain the status of the system, if necessary triggering off the protection system (partial or total cut-off or alarm, as necessary).
d) To control the process, i.e. to control the sequence of operations which the system has to carry out, having regard to the input data.
   The control unit stores all the data concerning the functioning of the photovoltaic field and the cold store, and it is accordingly able to optimize the operating functions of the system.
   Optimal functioning is achieved by controlling two parameters, namely the motor speed and the number of cylinders in operation in the compressor. The motor speed can be regulated by means of the excitation current, and by varying the number of cylinders in operation in the compressor it is possible to change the shaft resistance torque. There are thus two degrees of flexibility of control.

10.2 Control of the ozonizer system

The purposes of controlling the system are :
a) To produce analog signals to pilot the DC/DC converter by modifying the apparent impedence which the photovoltaic field follows to achieve MPT. Moreover, the DC/DC converter can achieve maximum power tracking by inter-acting with the control unit in terms of minor increases or decreases in the current and the monitoring of the resultant increase in power.
b) To control the correct battery charging and discharging cycle, by deviating from MPT logic when necessary.
c) To monitor the status of the system and, if necessary, to trigger off the protection system (partial or total cut-off or alarm, as necessary).
d) To monitor the sequences for switching the system on and off.

10.3 Data acquisition

In addition to controlling the ozonizer system, the VIDIMAC 8000 microprocessor also controls data acquisition for the system as a whole. It comprises the following units :
a) a central processing unit (CPU);
b) acquisition system for analog and digital data (DAS);
c) a VDU;
d) a printer;
e) a standard ECMA 46 cassette recorder, or alternatively a DEC (RX 02) compatible floppy-disk recorder;
f) a clock card.

In addition to the acquisition and monitoring system there is a no-break power system and a synoptic board for the entire plant piloted by the VIDIMAC 8000 with a mixed hardware and software system.
The system has 110 analog inputs and 100 digital inputs which can meet the needs of both the ozonizer system and the cold store.

The analog signals input to the DAS are standard DC 4 - 20 mA signals corresponding to the sensor's entire measuring field. These signals will be converted into 1 - 5 V, DC voltage signals as they are input into the DAS by means of 250 K Ohm precision resistance.

To avoid earthing ring currents interfering with the signal conductors and the VDU, there is to be a galvanic separation, on the sensor side, for all the signals and earthing of the VDU conductor at a single point near the sensors.

The digital inputs to the DAS are voltage-free contacts which the DAS (VIDIMAC) will supply at 24 V, DC.

Operation will be in negative logic, hence the closed contact will have the real value (T or "1") of the digital variable.

BIBLIOGRAPHY

(1) "Water disinfection system and cold store" - Final Design Review Meeting, Brussels 30 November - 2 December 1981.

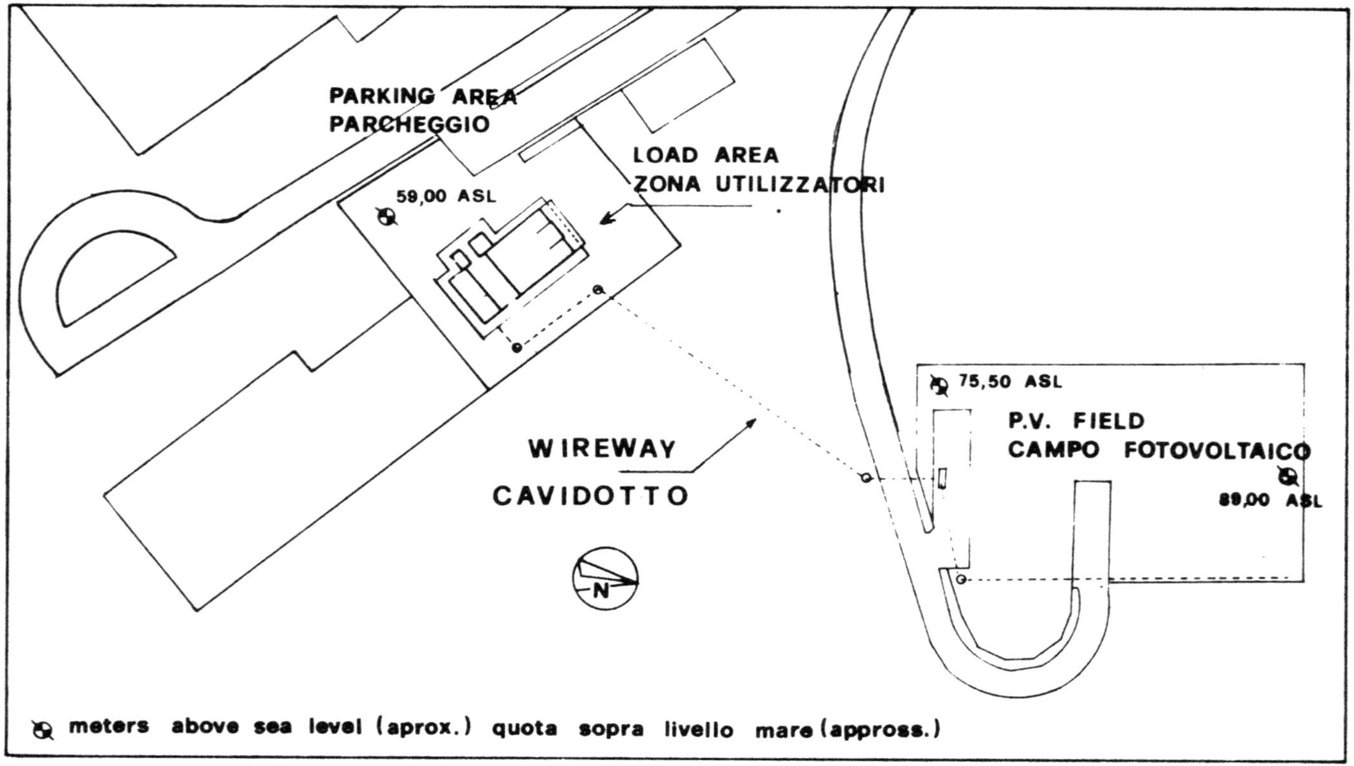

Fig. 1 - Key plan

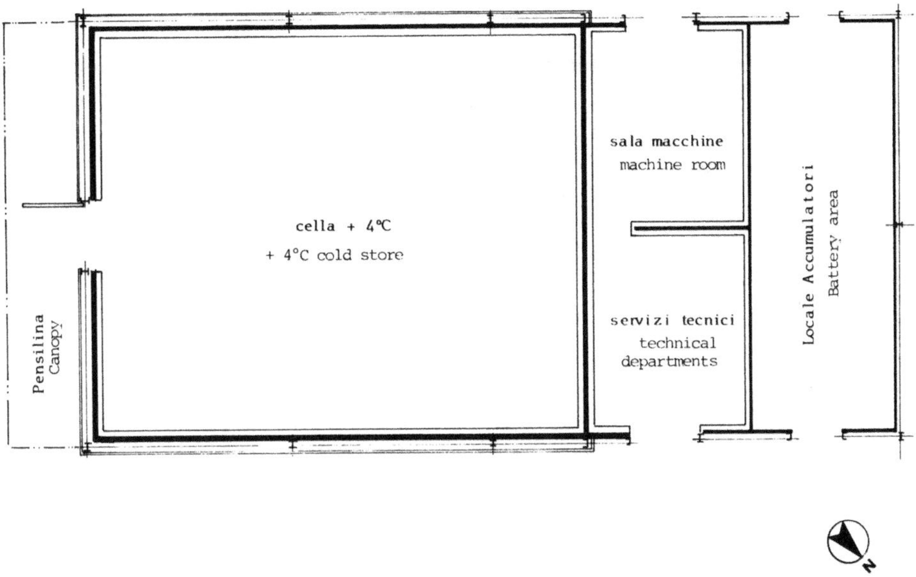

Figure 2 - Diagram of prefabricated building

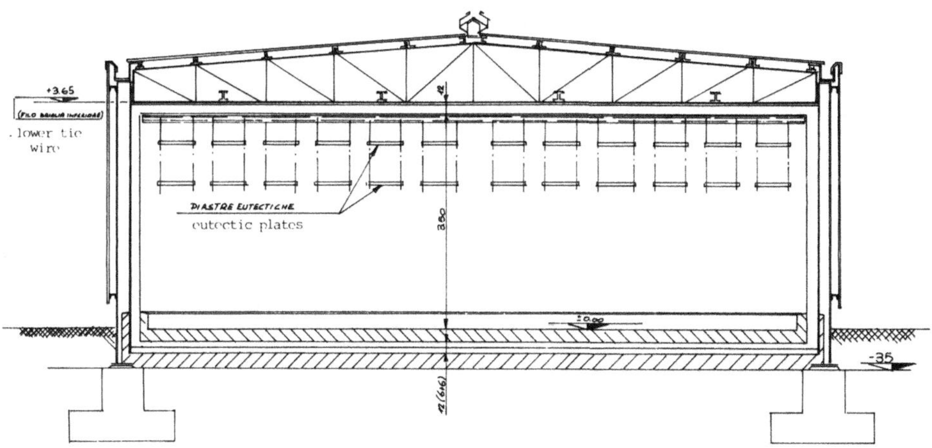

Figure 3 - Cross section of cold store

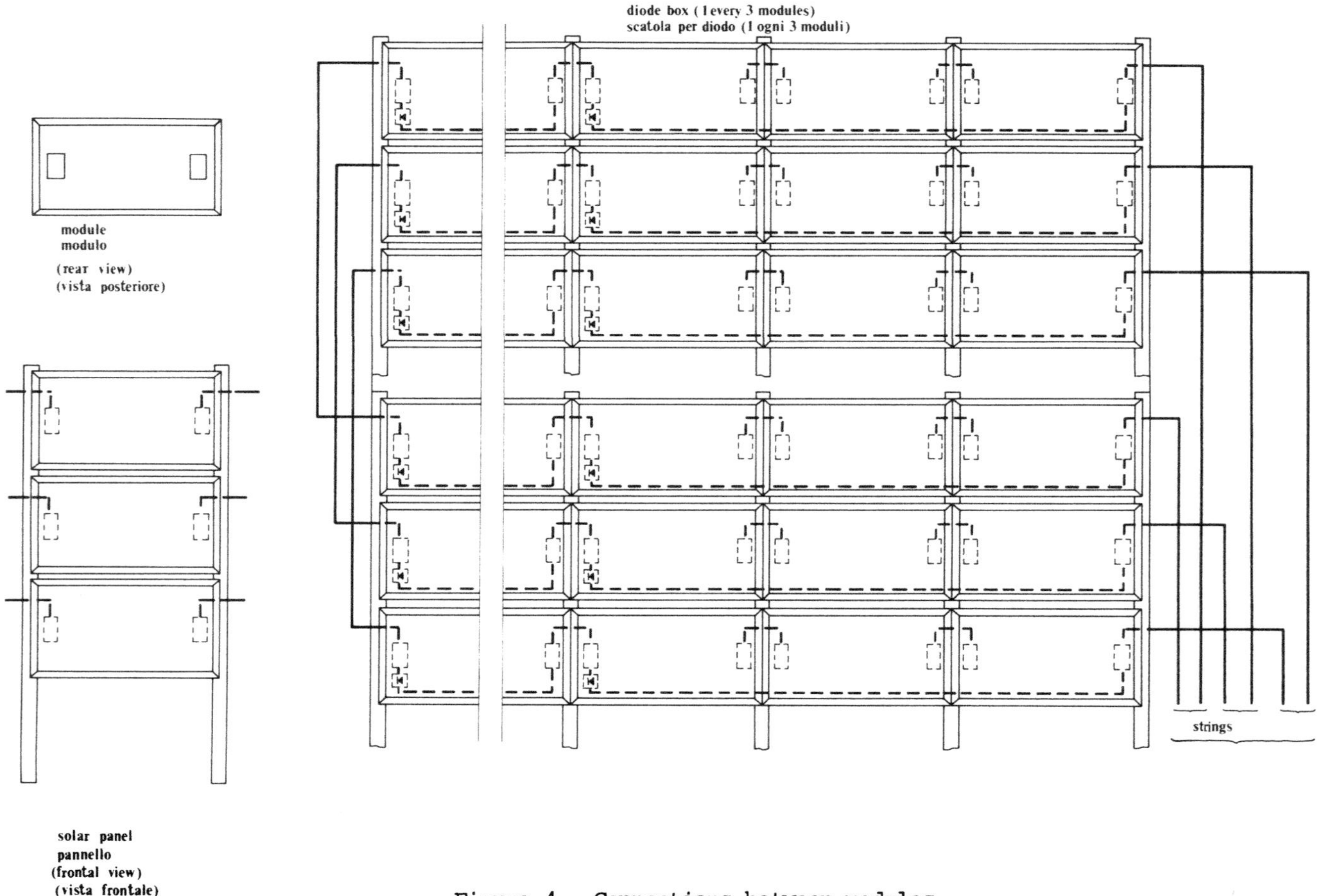

Figure 4 - Connections between modules

PASAN SUN SIMULATOR II

Panel : 72I'G/SOL
Number: 432196 9

Date : 7/3/83
Sign : M

1	2	3
Temp=0 °C	Temp=50 °C	Temp=25 °C
Pmax=61.499 W	Pmax=49.101 W	Pmax=55.43 W
Isc =13.261 A	Isc =13.849 A	Isc =13.555 A
Voc =7.4401 V	Voc =6.0895 V	Voc =6.7648 V
Eff =10.881 %	Eff =8.6874 %	Eff =9.8071 %
FF =62.332 %	FF =58.222 %	FF =60.449 %
Vpm =5.4516 V	Vpm =4.2619 V	Vpm =4.9372 V
Ipm =11.281 A	Ipm =11.521 A	Ipm =11.227 A

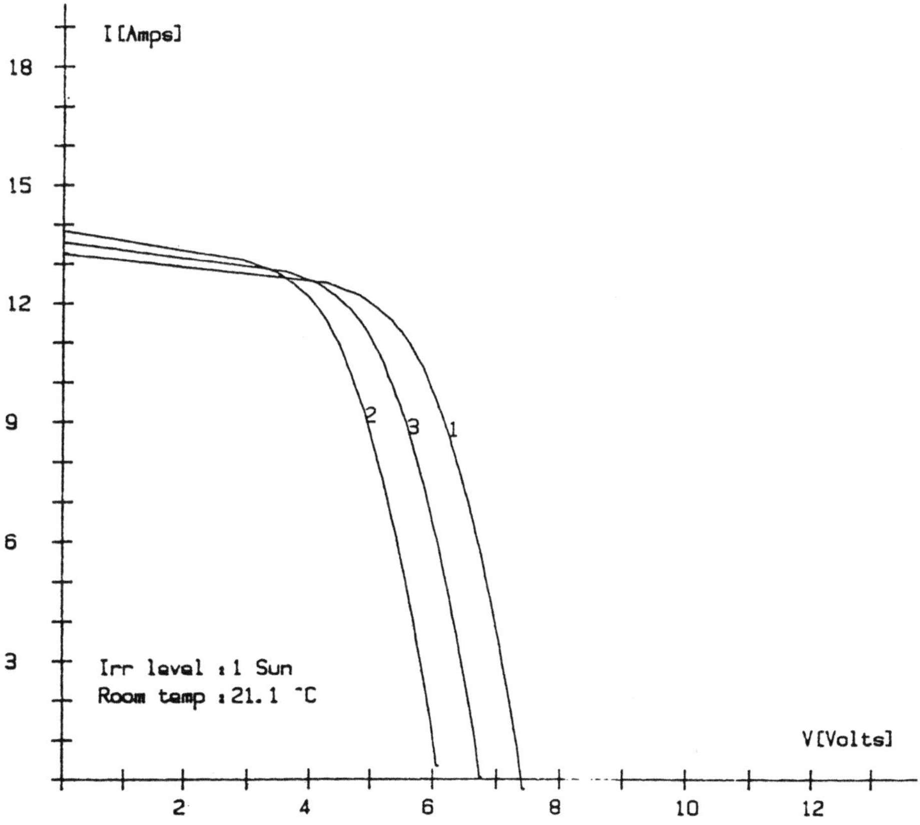

Fig. 5 - Specification of the panel

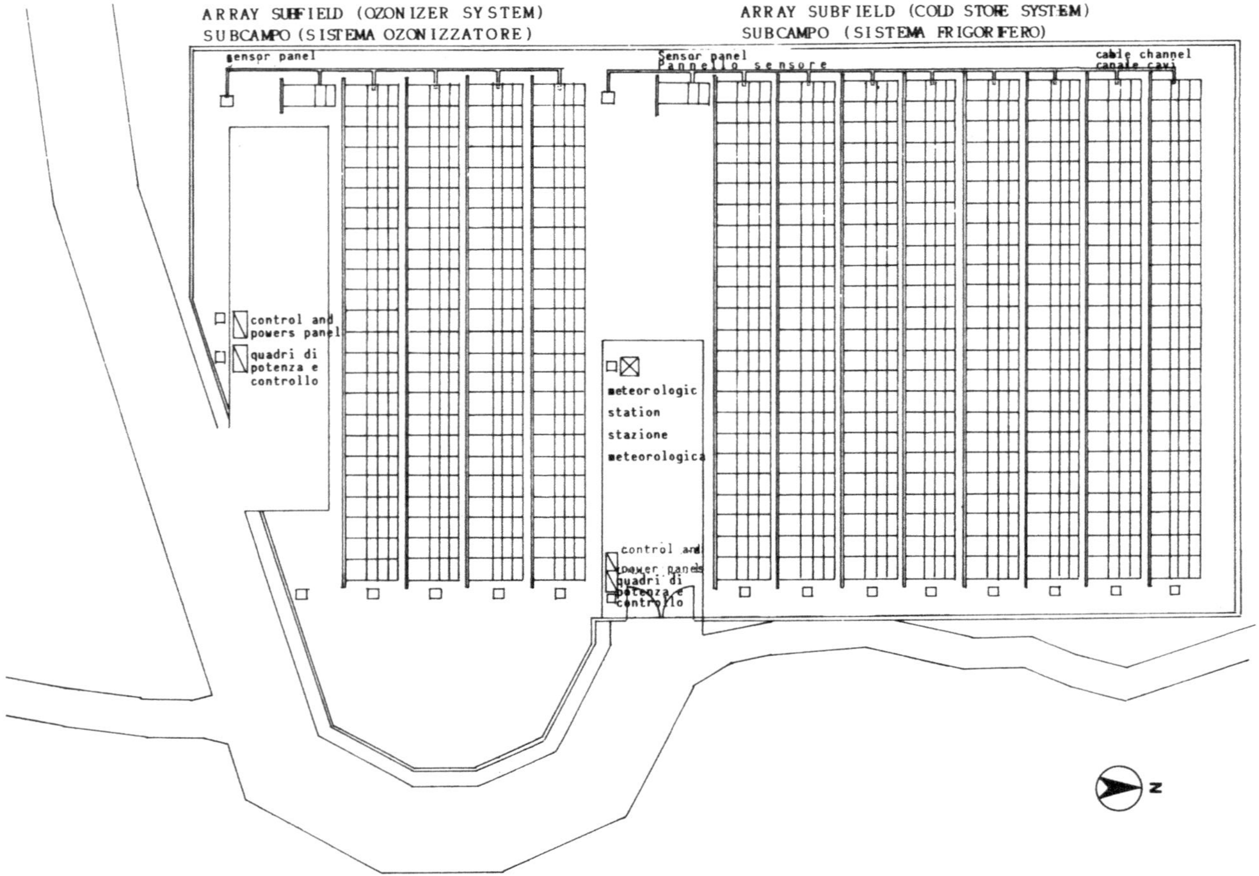

Figure 6 - Photovoltaic field

Figure 6 - Photovoltaic field

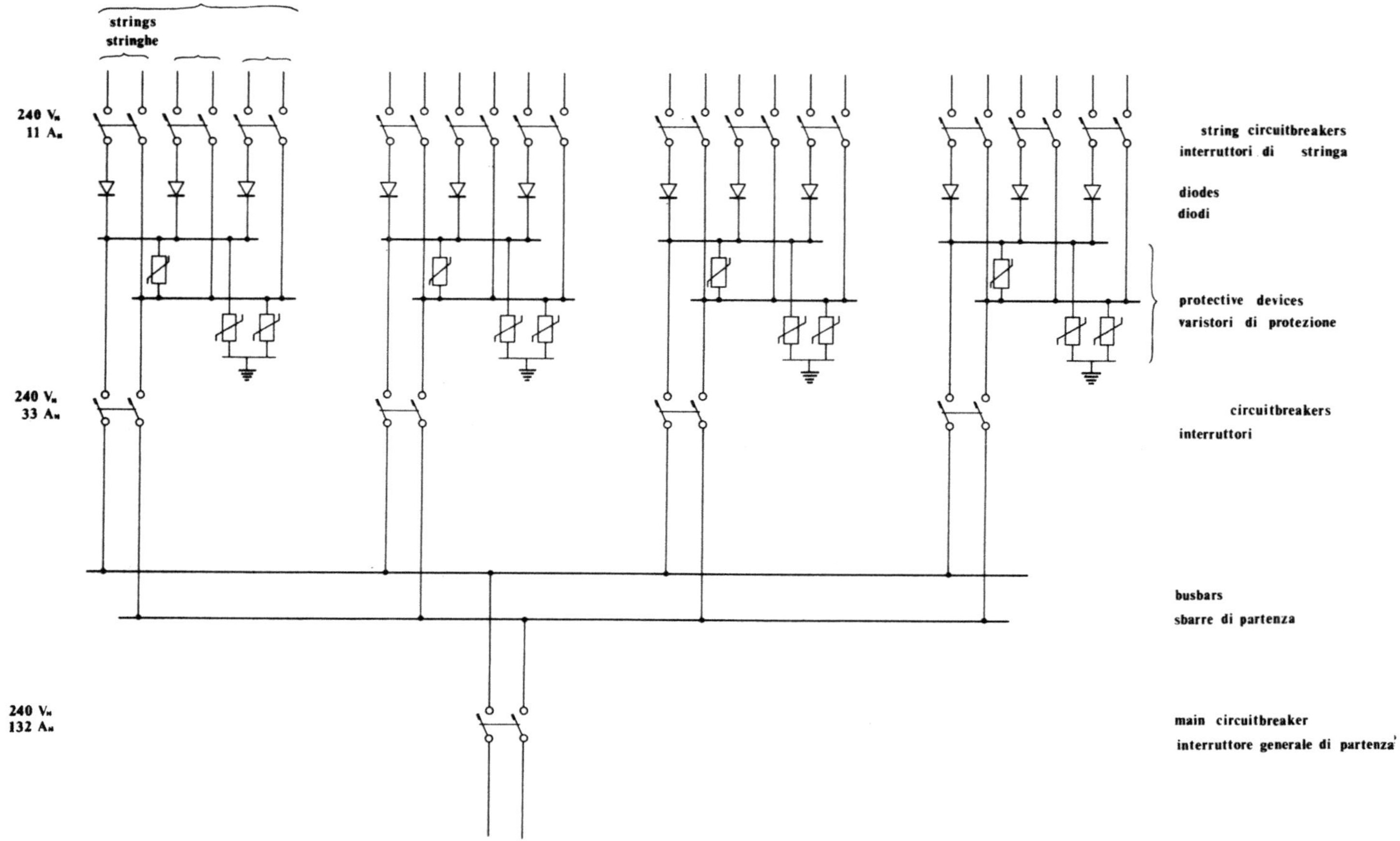

Figure 7 - Power panel board, cold store system

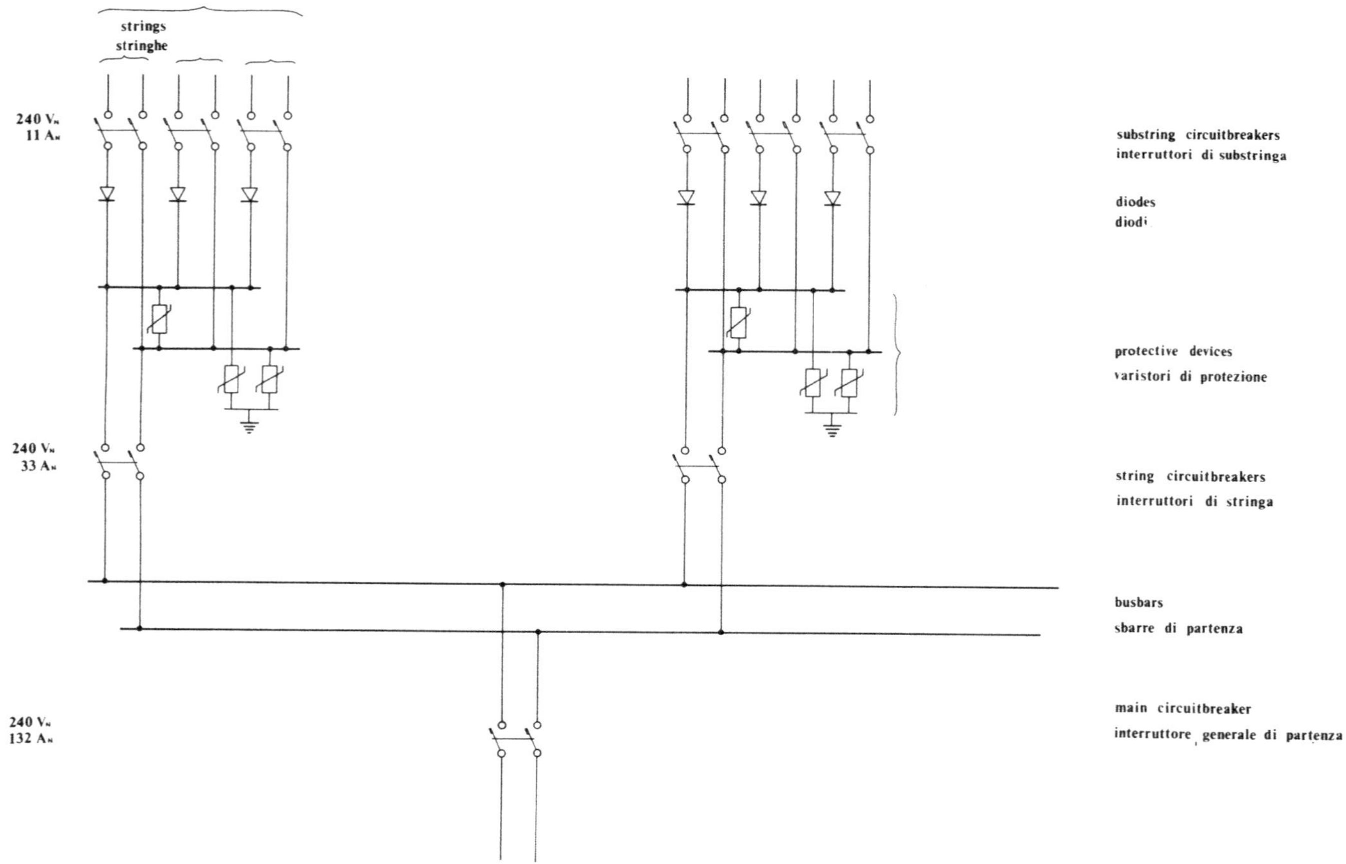

Figure 8 - Power panel board, ozonizer system

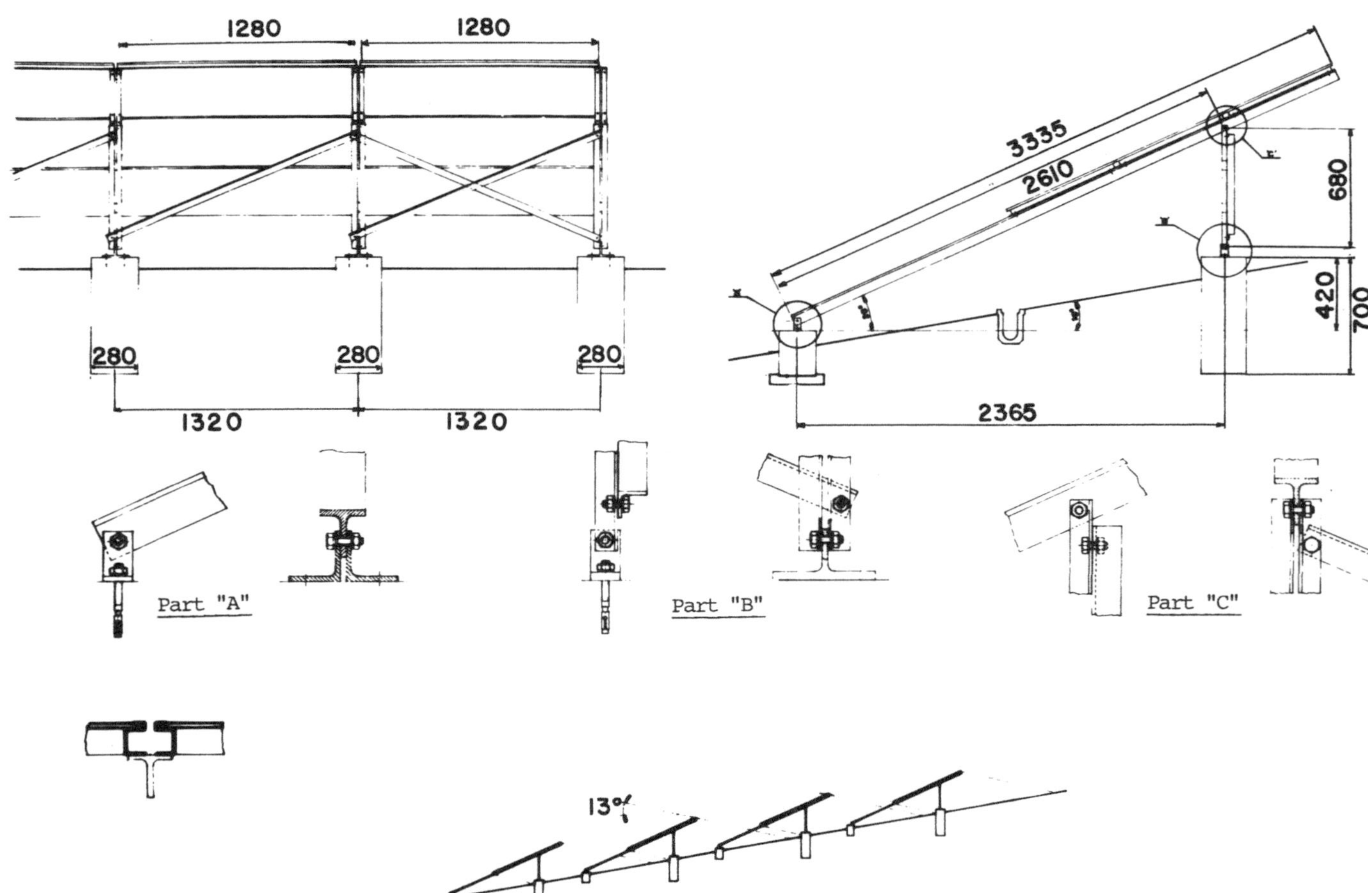

SCHEMA DI OMBREGGIATURA
Figure 9 - Supporting structure for panels

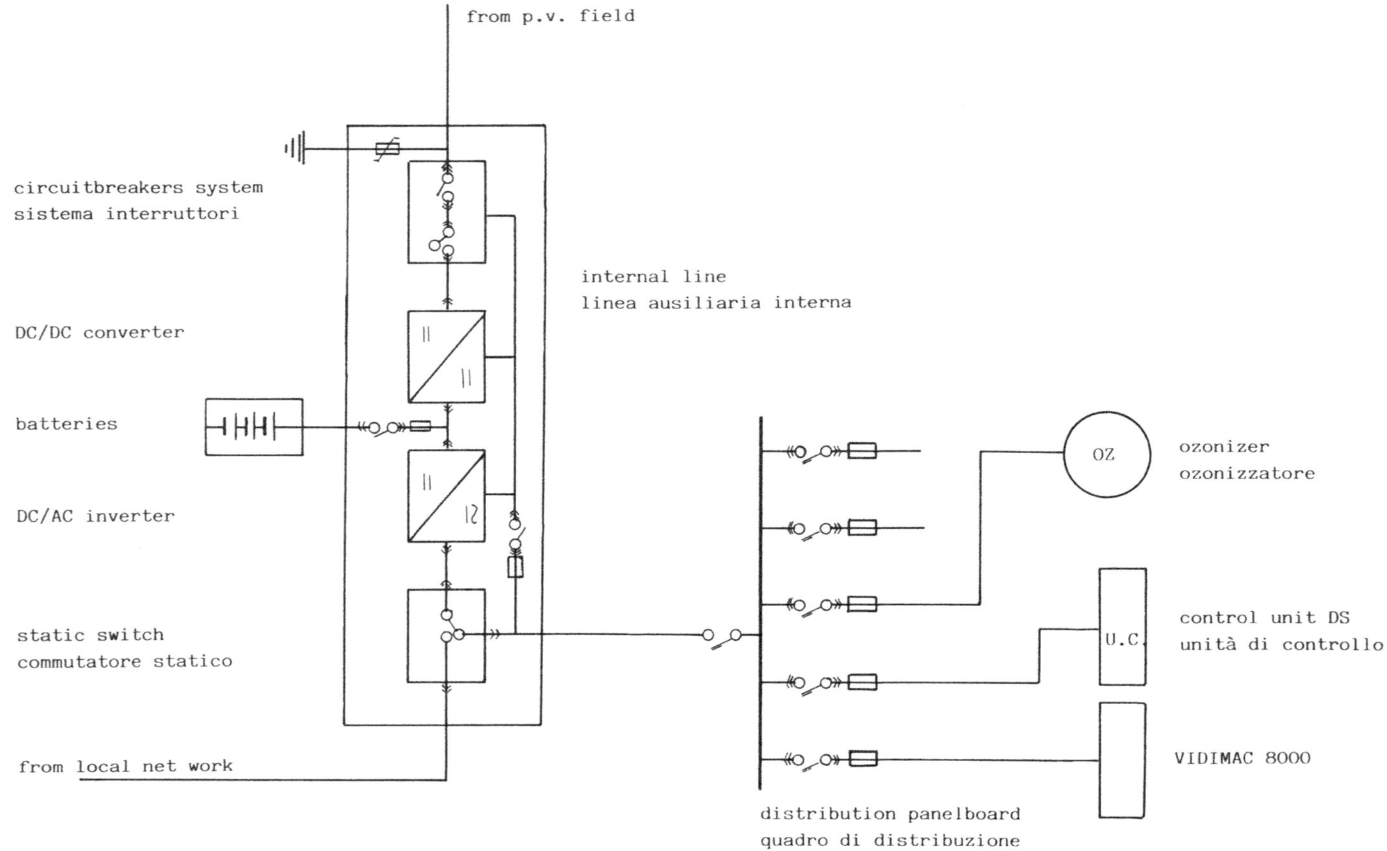

Figure 10 - Diagram of ozonizer power supply

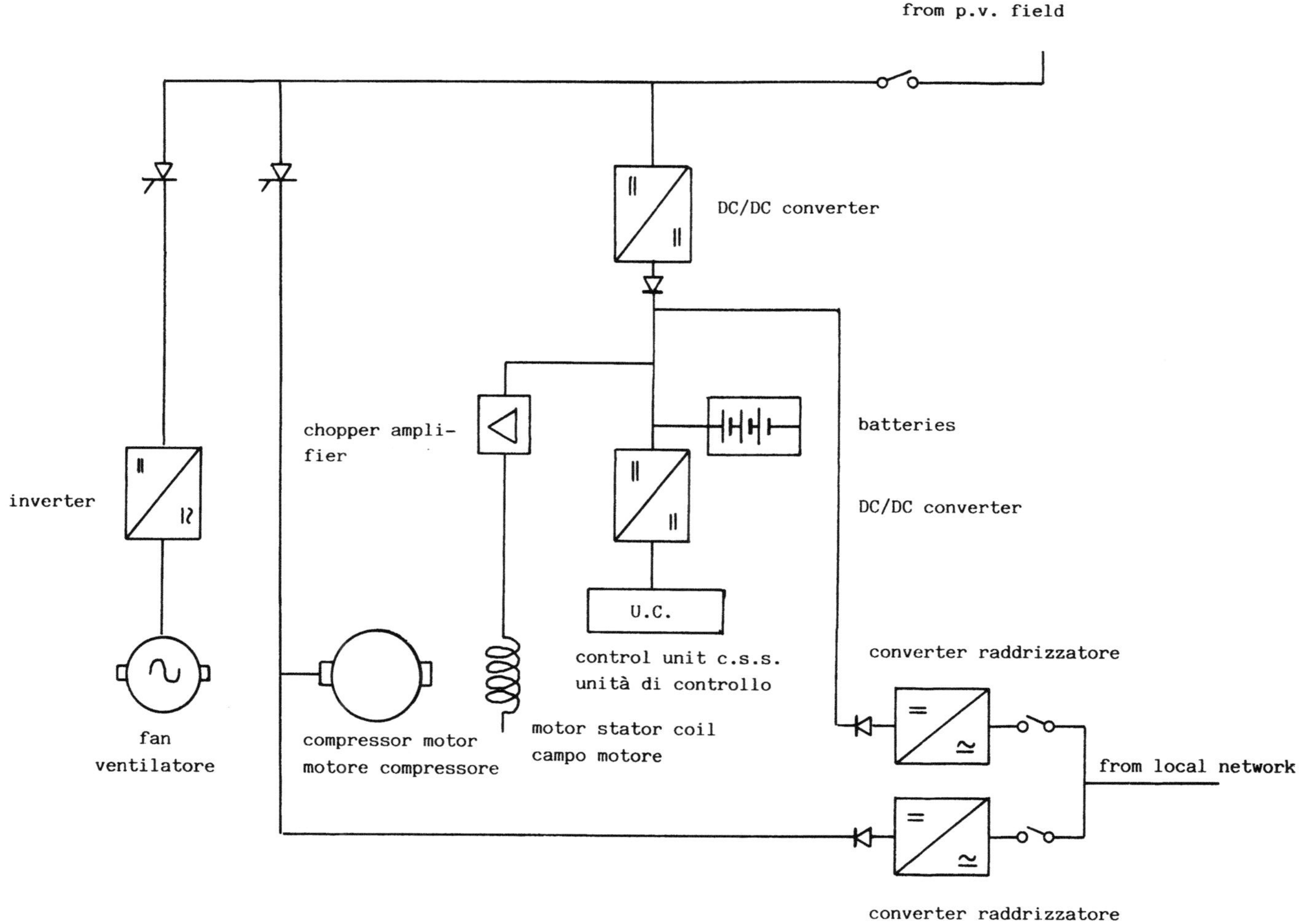

Figure 11 - Diagram of cold store power supply

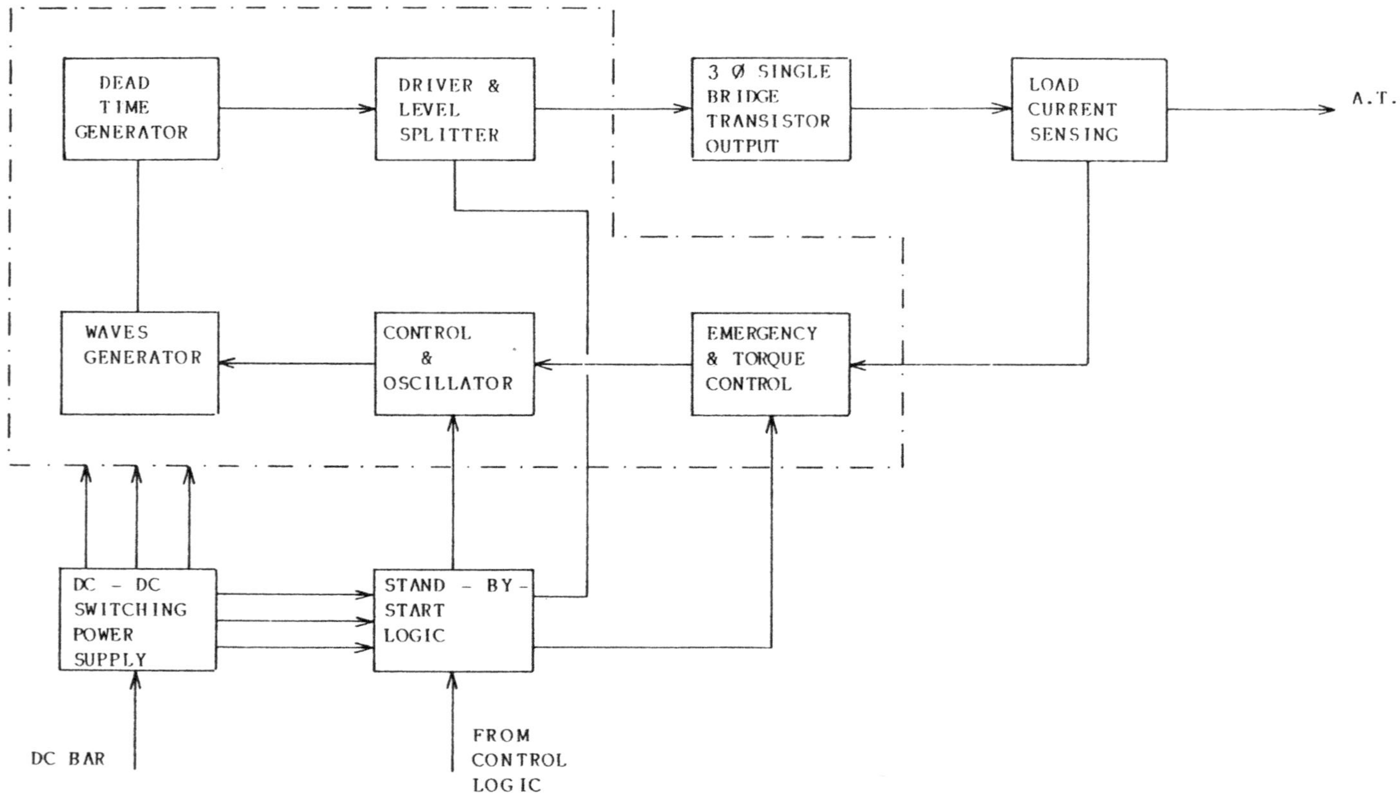

Figure 12 - 12 KVA inverter block diagram

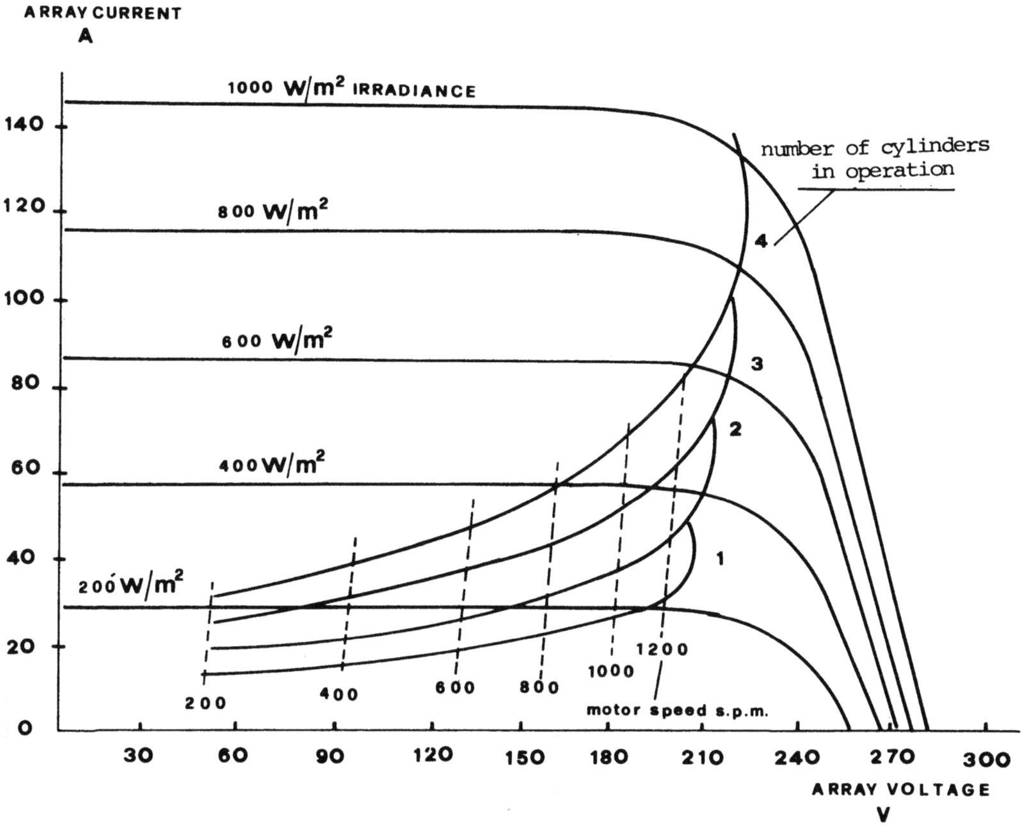

Figure 13 - Characteristics of matched systems motor/voltaic field

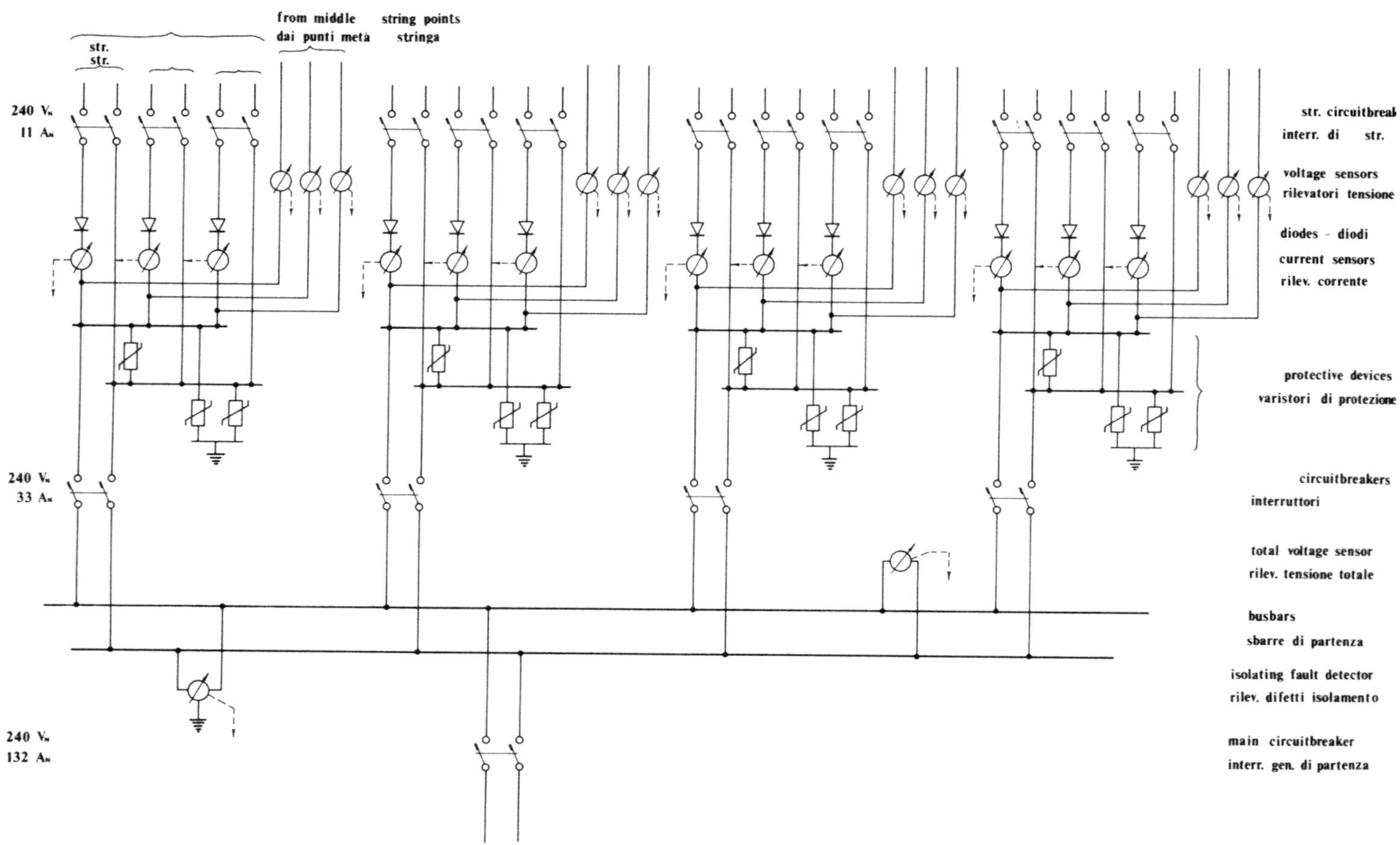

Figure 14 - Sensors, cold store system

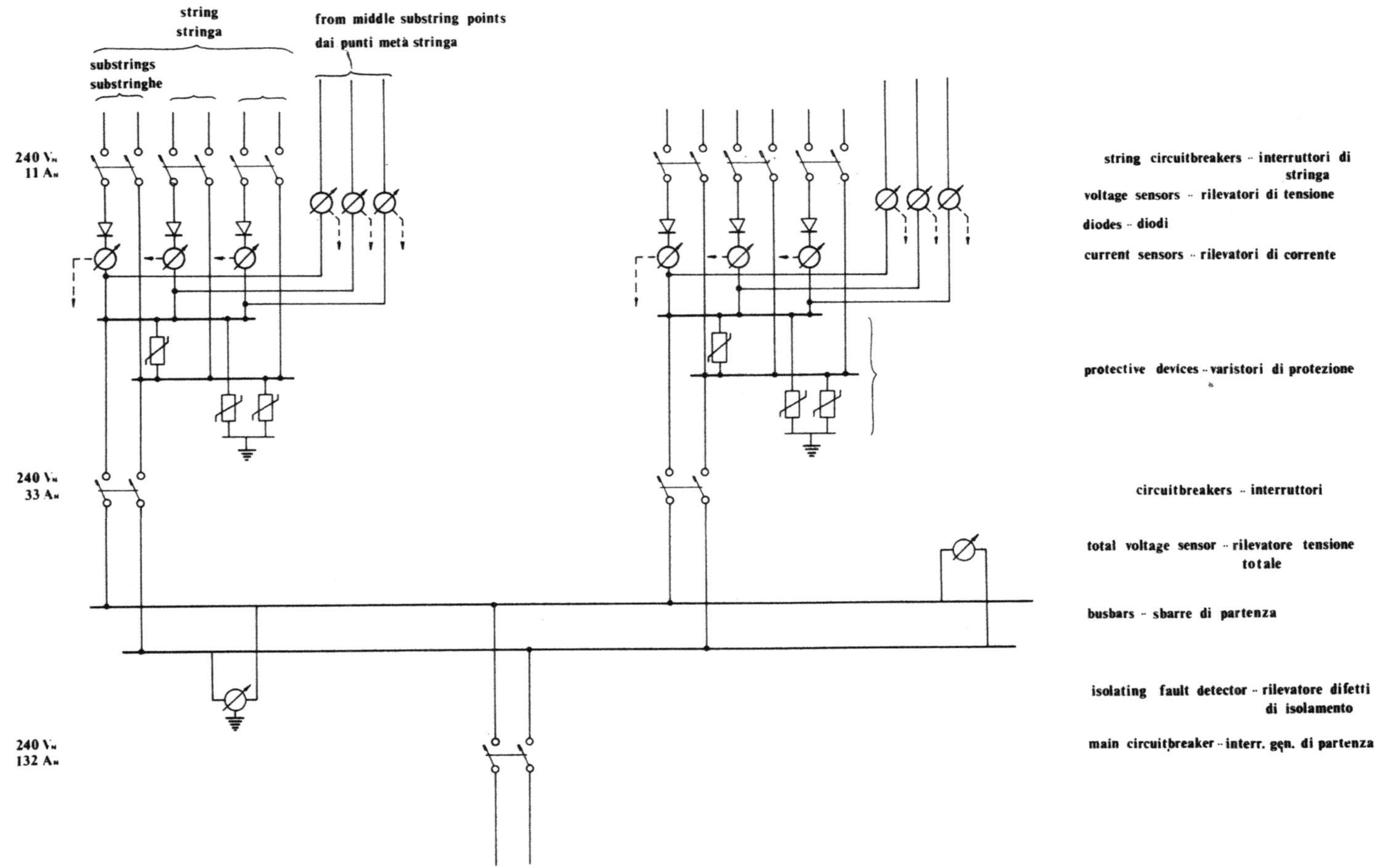

Figure 15 - Sensors, ozonizer system

SOLAR PLANT FOR A REMOTE CORSICAN VILLAGE
REVITALISATION D'UN VILLAGE CORSE

Location, country	: PAOMIA-RONDULINU, CORSE, FRANCE
Rated power of the array	: 44.064 Wp
Contract number	: ESL - P - 057 - F (D)
Duration of contract	: 27 months
Total budget of the project	: 948.750 ECU
E.C. contribution	: 308.000 ECU) 39 %
E.C. Bonus	: 61.500 ECU)
Name of other funding bodies and share of total cost	: - E.D.F. (ELECTRICITE DE FRANCE) 362.300 ECU (38 %) - A.F.M.E. (Agence Française pour la Maîtrise de l'Energie) 122.500 ECU (13 %) - MOTEURS LEROY-SOMER 94.250 ECU (10 %)
Contractor	: - MOTEURS LEROY-SOMER (F) - FRANCE-PHOTON (F) - OLDHAM BATTERIES LIMITED (GB)
Project Leader	: Dominique MERCIER MOTEURS LEROY-SOMER Usine des AGRIERS 16015 ANGOULEME Cedex FRANCE Tél. (45) 95.70.66 Télex 790244 F

ABSTRACT :

The PAOMIA Project is a stand alone photovoltaïc plant installed in a remote area in Corsica located 10 km from CARGESE. The aim of the plant is to power the RONDULINU hamlet and houses located in its area. This hamlet is now losing its inhabitants due to the lack of electric power. It is expected that the today's inhabitants will now stay there and that new people will come back to this marvellous site.
The plant will feed the houses, public lighting, a water pump. It is expected that a cottage industry and some touristic activity will start after electricity will be available.
The plant consists of a 44.064 Wp solar array installed nearby the hamlet. Power management includes a 50 kVA triphase PWM inverter, a battery bank of 84 cells of 2500 Ah rated capacity, a battery charge regulation and control and a measurement computer system. A back up 25 kVA generating set is provided to charge the battery in place of the solar array in case of very bad winter seasons.

EC PHOTOVOLTAIC PILOT PROJECTS — PAOMIA PROJECT — CORSICA — LEROY SOMER — FRANCE PHOTON — OLDHAM

I - INTRODUCTION :

The PAOMIA project has been designed as a stand alone facility to power the RONDULINU hamlet in the PAOMIA area near CARGESE (Corsica). This hamlet is located on the spurs of the mountains at about 400 m. height. No power is available on site and the first line of the grid is 4 km away. This hamlet was used by the inhabitants of the mountain's village of Vico as a winter place to feed their cattle and ewes on the winter's green meadows. This ancient habit is now near to be lost due to the lack of electricity on site and 7 families only are living there in winter. A few people born in this village are coming there in summer for hollidays.

The PAOMIA area is a very interesting place with marvellous sightseeing and countryside :

The site is in face of the CARGESE bay and oriented to the south and the remains of a 17th century greek village are close from the hamlet. However, no touristic activity was possible without energy.

The Syndicat d'Electrification de la Corse du Sud was very interested by a solar plant to supply power to this area. It has canceled a connection to the grid planed for 1985 and it will be the owner of the plant which will be operated by Electricité de France. The aims of the plant can be listed as

- Supply of 220/380 V. 50 Hz three phase power to a local grid, on a base as close as possible of the E.D.F. standard and habits,

- Connection of the houses of the hamlet and of the street lighting to this local grid,

- Give allowance for inhabitants to develop a cottage industry and/or touristic activity in summer,

- Power a submersible water pums (No water is available in the houses of the hamlet),

- To gather data and experience on how a stand alone solar plant can be used and managed to power small villages.

The choice of a stand alone facility is especially interesting for E.D.F. which can be involved in a lot of such projects in other remote area in FRANCE or in overseas French departments. To achieve this last goal, the PAOMIA location presents the following advantages :

- Typical consumer type load,
- No connection to the grid,
- Very sunny area,
- Easy to reach site (about 75 minutes from the AJACCIO international airport).

The design of the plant has been made to match the requirements of such an experiment :

- Photovoltaic array mounted on easy to tilt adjustable frames
- Large battery storage (7 days storage on the winter production base, 3 days on the summer production base),
- DC/AC solid-state transistorized self commutated inverter with a very flat efficiency v.s. power curve,
- Simple and reliable power management system,
- Back up generating set included to allow continuous power supply even in very bad weather conditions,
- Complete self operating measurement system with a long term data recording.

As this pilot plant is also designed to be a demonstration and a experimentation field for future plants to install in E C countries or to export, it has been designed and constructed using the simplest possible way. Standard components has been used where it was possible, and special parts has been designed as far as possible to have the possibility of transferring their manufacture in further projects, and to allow erection and mounting of the plant by local subcontractors.

The task repartition was the following :

- Project Management	: Moteurs LEROY-SOMER (ANGOULEME)
- P.V. modules	: FRANCE PHOTON (ANGOULEME)
- Batteries	: OLDHAM BATTERIES Limited (UNITED KINGDOM)
- Inverter	: Société Nationale AEROSPATIALE (TOULOUSE)
- Generating set	: C.L.M. PEUGEOT (PARIS)
- Measurement system	: PME INFORMATIQUE (ANGOULEME)
- Surge protection	: SOULE (BAGNERES-de-BIGORRE)
- Power management	: LEROY-SOMER (ANGOULEME)
- Frames	: LEROY-SOMER (ANGOULEME)
- Wiring	: ENCO RAFFALI (AJACCIO)
- Civil work	: A. BOUZIANE (AJACCIO)
- Architecture	: G. FELIX (AJACCIO)

General design and other parts has been made by LEROY-SOMER (ANGOULEME).

The plant will be operated by ELECTRICITE DE FRANCE (E.D.F.) under the responsability of the SYNDICAT D'ELECTRIFICATION DE LA CORSE DU SUD (Ajaccio) which will also pay and install the distribution nettwork.

2 - GENERAL TECHNICAL DESCRIPTION OF THE PROJECT :

2.1 - SITE :

The PAOMIA village is located in the mountains near Cargese surrounding the Cargese bay.

The Rondulinu hamlet is about 5,5 km from Cargese at an altitude of about 440 m. Cargese itself is about 35 km from Ajaccio (airport). The center of the hamlet is near a chapel and includes 5 houses out of which 3 are permanently inhabited. Two other parts will be fed : 4 houses 300 meters north of the center and 4 houses, 350 m south west of the center, just as 5 to 10 isolated houses in a 500 m. radius from the plant.
It is surrounded by 2 hills of 635 m. and 549 m. the top of which are at 750 and 500 m. from the plant.

The plot itself is located on a very tilted hill south of the hamlet. It has two main slopes : 20 degrees from North to South, 8 degrees from East to West and is oriented south west-wards (30 to 40 degrees).

This lead to divide the array in small rows installed on horizontal platforms distributed in the plot.
The array is oriented southwest ward 20 degrees from the south. This choice comes from :

- The shadow due to the adjacent mountain chain in the morning in winter,

- The small effect of this orientation on the energy production of the plant (less than 4 % without East shadowing),

- The minimization of civil work.

The technical building is located between the solar array and the plant and is about 15 m. away from the first house of the hamlet.

2.2 - CLIMATIC CONDITIONS :

We have no solar radiation recording available for PAOMIA or Cargese but a complete set of radiation data are

available for Ajaccio (26 km in a bee's line). We have used this recording and correct it to take into account the latitude and altitude variation. This measurement has been made by Meteorologie Nationale (Table 1 - Climatic conditions).
The plant has been designed using a LEROY-SOMER computer program to calculate the energy received on a average basis on tilted panels.
The non-solar climatic data used for the design of structures has been found in French building standards as follows (height and shape corrections made) :

Nominal wind load : 81 da N/m2 (131 km/h)

Maximum wind load : 141 da N/m2 (173 km/h)

Nominal snow load : 17 da N/m2

Maximum snow load : 24 da N/m2.

The correction has been made taking into account an array height of 5 m. (actual height : 2 m.), a tilt angle of 60 degrees and an exposed site Réf : DTU NV 65.

	Insolation ground Wh/m2/day	Insolation panels at 60 degrees southwards Wh/m2/day	Insolation panels at 60 degrees 20 degrees west wards Wh/m /day
January	1849	3363	3236
February	2766	4246	4102
March	3624	4300	4171
April	4994	4797	4615
May	6044	5022	4842
June	7067	5343	5201
July	7370	5711	5511
August	6240	5594	5414
September	4849	5334	5122
October	3492	4923	4757
November	2139	3685	3551
December	1721	3458	3319

2.3 - DESIGN STRATEGY :

- The sizing of the array has been made taking into account the average climatic condition and using average efficiencies for all components especially batteries and inverter.
- The sizing of the batteries has been made to insure a storage duration sufficient to go over the medium estimated variations of the solar energy received in winter. A 7 days storage on a winter basis will give a sufficient security for most of the years. Due to the summer increase in production and consumption, this storage will give a 3 days reserve for July and August.
- The sizing of the inverter results of two opposite choices :
  . The larger the inverter, the smaller the number of automatic disconnections due to the starting of motors..., and the easier can the increase in demand be met,
  . The closer the power of the inverter to the peak power demand the higher is the global efficiency. The chosen inverter power of 50 kVA allows a maximum load of 62,5 kVA (see below) and then allows in the worst case (August, 46 kVA) to start motors of about 8 kVA (starting factor of 3),
- The sizing of the diesel generating set has been made to ensure a complete recharge of the batteries in less than two days. It will start when the battery charge level is below 20 % of the nominal charge and will be connected directly to the batteries throught the charge regulator and in parallel to the solar array to avoid any loss of solar energy.
- The array is divided in medium size subarrays and each subarray is directly connected to the general power management device in the control rooms. Therefore, it will possible to monitor each subarray characteristics and to detect a module failure (see below).

3 - ARCHITECTURAL AND ENVIRONMENTAL ASPECTS :

The PAOMIA site is located in a "sensitive area" and then it has been difficult to get the building agreement for the plant. The architectural work has been made under the direction of a local architect and care has been taken about the aspect of the plant :

- Divided array with small rows following the slope of the plot,
- Small technical building,
- Technical building and platform's supporting walls coated with a parget made with local sandstone and leaved with a rough aspect,
- Traditional fashioned doors and shutters made with local pin.

It has to be noted for further projects that the time needed for the building agreement was larger than any component delivery time (inverter excepted), or than the civil work construction time.

The work for civil work was a lot larger that expected at the begining of the project, due to :

- The need to erect platform to avoid erosion during the heavy rains occuring on site,
- The heavy difficulties to use civil machines due to the slope of the site. Almost all of thecivil work has been made by hand with shovels...

The plant can be seen from the track leading to the village and seems to fit correctly in the site.
It is also visible from the seaside road but is to far from it to disturb any thing in the landscape.

4 - OVERVIEW OF NEW TECHNOLOGICAL DEVELOPMENTS :

This project leads to new developments in products and engineering :

- Supporting frames have been developed for this particular project and are now proposed for other projects.

They have been designed as a selfsupporting structure with a high tilt adjustement versatility and as simple as possible : they are made with only three differents parts and are very easy to manufacture every where,

- The inverter has been studied and designed especially for the pilot project program by Aerospatiale,

- The power management system has been improved during the project,

- The measurement system has been designed for this project using a standard computer base and will be used as a model for further projects.

5 - PHOTOVOLTAIC MODULES :

The FRANCE-PHOTON modules used for this plant have the following characteristics :

. Rated peak Power	: 36 Wp
. Size	: 1281 x 642 x 42,5 mm
. Weight	: 14 kg
. Number of cells	: 36
. Nominal voltage	: 13 volts (21,5 V open circuit voltage)
. Cells	: 4 inches monocristalline silicon cells
. Encapsulation	: . tempered low iron glass . transparent silicone rubber . white silicone rubber.

The modules has been mounted in the FRANCE-PHOTON Workshops in 6 modules panels using a light L shape frame.
One single connection box is used for this 216 Wp panel to minimize misconnection problems (mistakes, corrosion,...)
The 6 modules are serie-connected in the connection box by FRANCE-PHOTON.

The modules have been sorted by FRANCE-PHOTON on a current base. After sorting, the mounting of the panels has been made in chosing the same current class for the twelve modules needed to build 2 panels having the same rated current. This two panels have been serie connected together on site to reach the nominal voltage of the array with the minimum mismatch losses. The work on site has been reduced a lot by packing the panels needed for a single row (i.e. 1296 Wp) in the same box with labels to identity the matched panels.

A statistic study on the PV modules of PAOMIA was conducted. Results are as follow :

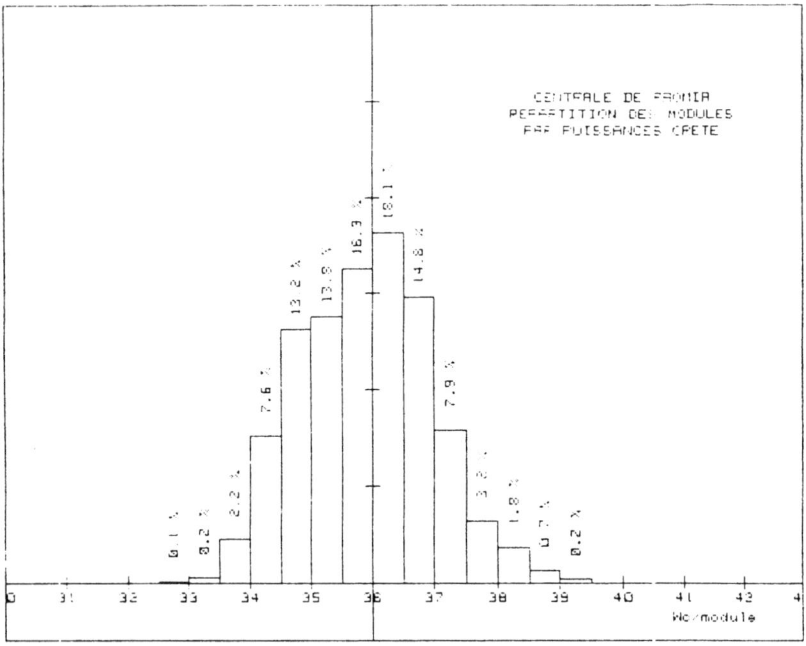

Peak Power histogram
Table II

Average power : 36 Wp ± 0,5 %
Standard deviation : 1.05 Wp

On an average base, in using this way for matching of the modules in each serie, the mismatch losses are less than 2 %.

6 - ARRAY AND WIRING :

6.1 - ARRAY SUPPORTING STRUCTURES :

The supporting structures of the modules is of a triangular shape with two identical posts without any other pieces exept bolts, ... It is made in galvanized steel and designed to support the highest wind and snow conditions of the site and to achieve a minimum weight by Wp.

This structure allows a tilt adjustment from -10 degrees north to the nominal 60 degrees of the plant's design.

The installation of the structure on the ground is very easy : one must only seal bolts into concrete benches after completing the mounting. The large base of the structure gives a very good wind resistance with a small amount of concrete.

The panels incorporating 6 in-fabrica mounted modules are directly mounted on the structure. The connection are made with a precabled connection box supporting the wires needed to connect it to the panels.

The sizing of this structure has been performed following the french standard CM 66 for calculation of metal structures. The thermostress are taken into account by soft plastic washers.

It must be pointed out that the structure has been designed for an easy erection without special machinery in the whole world and with a minimum number of pieces (only 2) in order to avoid losses and mounting mistakes.

6.2 - ARRAY OUTLINE AND WIRING :

The solar array is divided into 34 rows build with 6 six-modules panels. The rows are grouped in 5 parts distributed on the slope (see picture).

Given that each row has 6 six-modules panels, the wiring is made as follow :

- Each group of two six-module panel is connected to a single connection box (A) located on the frame between this two panels. The serie connection of the two panels is made at this level.

- The three series of each row are connected in parallel by wires leading the current from the (A) connection boxes to a two rows connection box (B).

- Two rows are paralleled in the (B) connection box.

- Each (B) connection box (17 (B) boxes for the whole array) is directly coupled to the power management system.

This wiring policy allow a very easy array trouble shoting :

- each two-row group connected to the power management system is made with 6 series. A simple current measurement made in the technical building can show if a two-row group is lower than the others, wich means that 1/6 of it (17 %) is over (i.e. one module is broken).

- A single measurement in the (B) connection box showswhich row is under performances.

- A check of the 3 (A) connection boxes shows which serie has problems.

- A check of the two panel's connection box leads to the broken module.

A general rake shape grounding wiring is provided to protect the array. Each frame is connected to this wires.

7 - LIGHTNING PROTECTION :

The keraunic level of Ajaccio is 37 thunderstorm days/ year) and then the earth flash density is about 2,5 flashes/km2/year.
Taking into account a surface of 1560 m2 without any influence of local conditions (horizontal array located in the middle of a hill) it means a probability of 4 10^{-3} direct strokes per year. One can also note that the top of the village's houses is located at a distance of meters of the top of the array and are about 15 meters higher...
Then, there is no protections against direct strokes on the array or the building.

However, the local utility will protect the 380 V distribution line (total about 1000 m) with the classical devices used for this kind of installation.

The protection against indirect effects of lightings has been made as follow :

. Equalization of the potential of the structure by interconnection of the supports of the panels to the general grounding of the technical building.
This grounding as well as possible to give the maximum safety level. A unique grounding will avoid a possible voltage difference between two grounds.

. Interconnection of all ground out-put of the devices included in the control building to the general ground of the plant.

. Complete isolation of the two polarities of the power circuit in respect to the ground.

. Installation of the different cables as close of each other as possible to minimize the surface of the induction loops.

Protection devices including diodes, MOV'S and coils have been installed at different levels of the array :

(A) connection boxes
(B) connection boxes.

The protection of the output of the inverter is under responsability of the distribution network.

8 - POWER CONDITIONING :

8.1 - CHARGE/DISCHARGE REGULATOR :

The output of the array consits of 17 pairs of wires arriving to the general bus bars of the technical building. After paralleling this wires, the power goes to the batteries while controlled by a charge regulator featuring :

. input current (from the array) : 350 A
. nominal voltage : 168 V
. overcharge cutt-off voltage : 201,6 V
. overdischarge cutt-off voltage (putting the generating set in operation) : 155,4 V
. output current (to the invertor): 375 A

The control cabinet features also the possibility of testing the open circuit voltage and short circuit current of each of the 17 subarrays.

8.2 - BATTERIES :

The energy storage of the plant consists of a lead acid battery bank.
This battery hasbeen manufactured following the solar requirement.
Its characteristics are :

- Type : OLDHAM SZT
- Capacity : 2500 Ah for a 10 h. discharge rate
- Cell nominal-voltage : 2 V
- Number of cells : 84
- Nominal bank voltage : 168
- Size per cell (LXwXH) : 212 x 576 x 820 mm
- Weight per cell : 191 kgs (electrolyte included)
- Plates : positive plate : tubular type
  negative plate : flat plate type
- Material : very low level of antimony lead alloy

- Self discharge rate : max. 4 % per month at 20°C
- Max. charging bank voltage : 201,6 V
- Max. discharging bank voltage : 155,4 V (80 % rate)
- Total storage capacity :
  415 kWh at a 180 A discharge rate
  560 kWh at a 60 A discharge rate.

The battery room is 3,5 m. height and features holes in the walls and chimneys to allow self ventilation.

8.3 - INVERTER :

The DC-AC inverter is of the PWM (pulse width modulation) type, entirely built with transistorized circuits.

In fact, the 3 phase inverter consists of 3 starconnected monophase inverters and allows a perfect phase equilibrium for any type of load. Each monophase inverter has a built-in filter commutation at a 15 % nominal power level to avoid excessive no-load losses. A galvanic insulation betwenn the three phases is done by an output transformer.

Characteristics :

. Nominal power : 50 kVA - 3 phase 220/380 V, 50Hz
. Load power factor : from 0 to 1
. Overall efficiency at 156 V battery (i.e. when the battery is close to be discharged) :
  5 kVA : 90 %
  10 kVA : 88 %
  25 kVA : 91 %
  50 kVA : 93 %.

8.4 - GENERATING SET :

The generating set is made with agas engine, and an alternator fitted with special regulation.
It has a nominal power output of 25 kVA and is used to charge the batteries in case of a sun failure in winter. The alternator is regulated to give a constant current at any voltage under its nominal voltage.
The current is connected to D.C. current after passing

through a transformer and then connected directly to the battery in parallel with the solar array. Using this way, there is no difference for the power managment system between the generating set and the solar array and it is possible to :

- store any current produced by the solar array during the running of the set,
- avoid any threshold problem in charging the battery with two sources.

The set is automatically started at the lowest level of the batteries and stopped at their maximum charge level .It will run for a full 2 days and, in case of a total lack of sun, stop 7 days if the consumption is at its winter level.
This delay will also allow time to do whatever will be necessary if the lack of power coming from the array is due to technical reasons.

9 - MONITORING SYSTEM :

The monitoring system has been made following the EC board requirements. It features :

- A standard microcomputer,
- A tape recorder,
- A very simple control board,
- A video screen operating on demand,
- A printer operating on demand.

The system has been designed to monitor each relevant parameter and to store them on a hourly base on a tape which can last about 4 months (ECAA 46 type).

- In normal operation, it runs automatically without screen to reduce energy consumption. Relevant parameters are printed each hour (one page per day).

- In human operation, one can monitor each parameter of the plant on the screen and print whatever he wants.

  The measurement systems records :

  - Meteorologicaland ambiant parameters,
  - Supporting frame stresses,
  - D.C. electric parameters,
  - A.C. electric parameters.

10 - SYSTEM INTEGRATION AND INSTALLATION :

No particular problem has been encountered in the installation of the plant. The only problem was that the on site work started on March 15, 1983 after finally getting the building agreement.
The plant was technically completed on June 30, 1983.

The wiring, the installation of the devices in the technical building last only two weeks and has been mainly done by local subcontractor. This proves at least that the design of the plant was sufficiently simple to allow quick installation by standard-skilled electricians...

11 - CONCLUSIONS :

This plant will be a model for future projects in remote areas and will provide electricity to a village which was dying before its completion. It proves that solar energy can solve energy problems for powering remote villages without special technical environment.

The Cargese Bay

The Rondulinu hamlet

The plot before plant installation

The plot before plant installation

50 KVA INVERTER

CIVIL WORK

Battery Bank

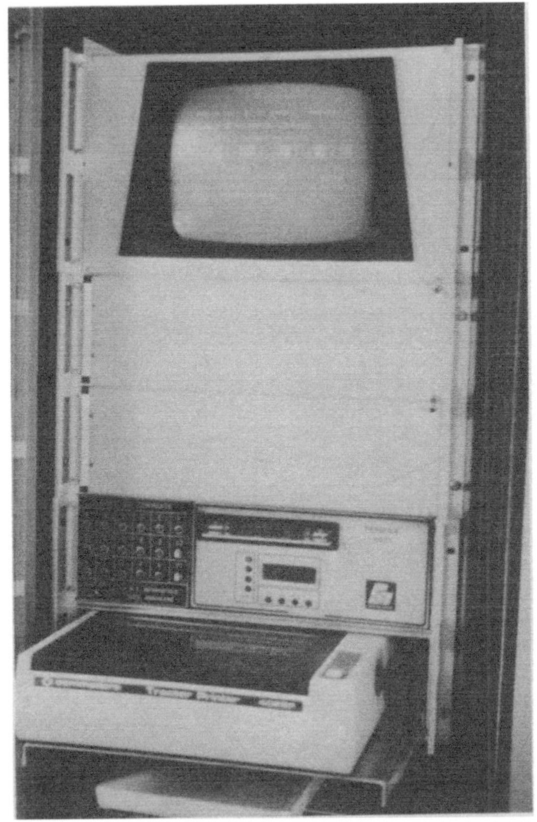

Measurement system

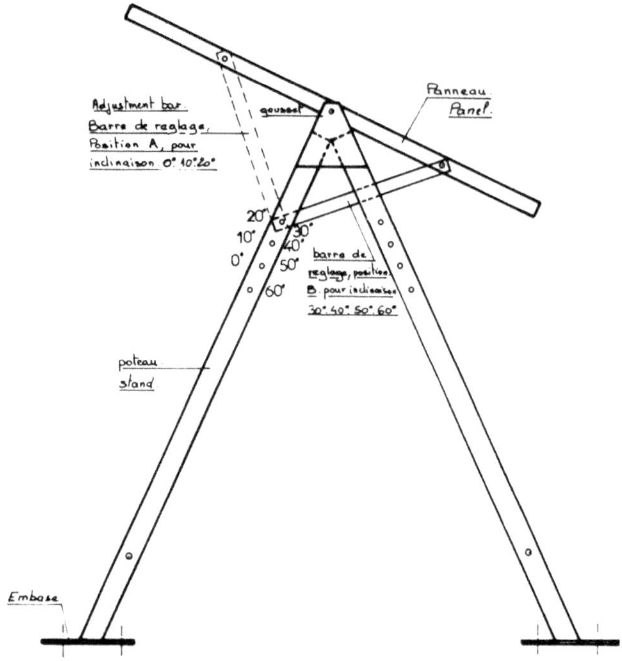

Schematic of frame adjustment

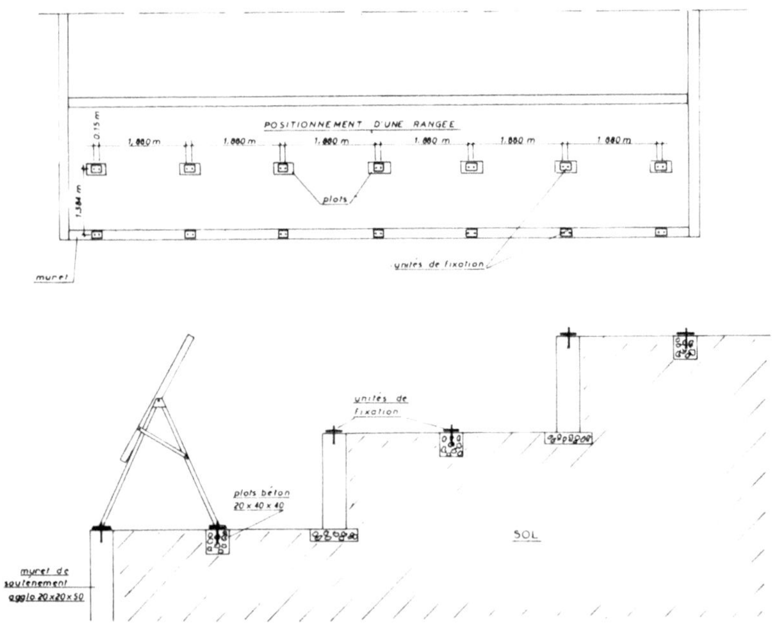

Supporting frame installation

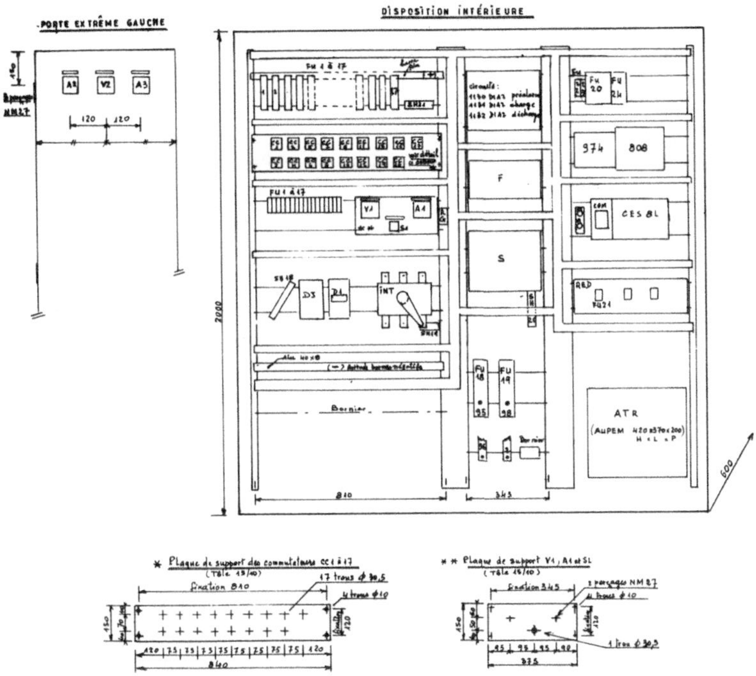

Power management cabinet

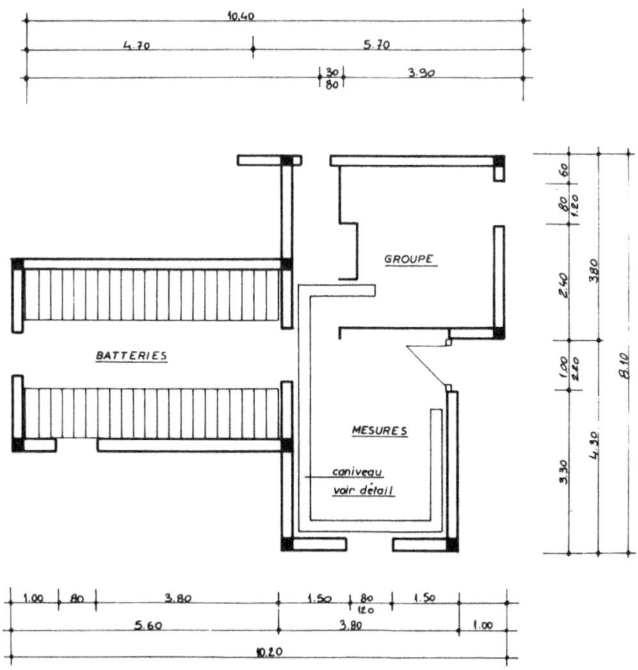

Technical Building

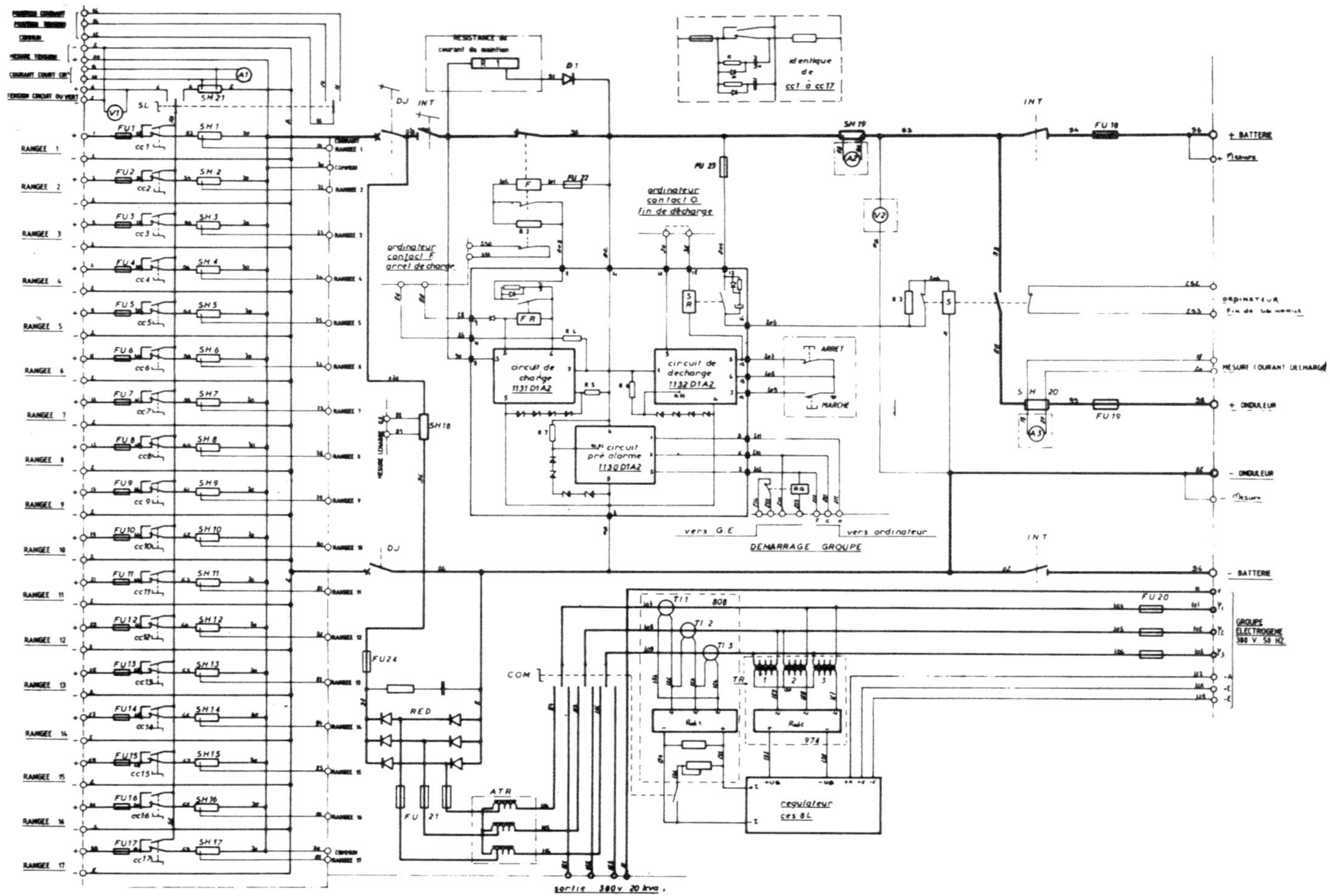

Power Management System Schematic

PHOTOVOLTAIC POWERPLANT FOR HYDROGEN PRODUCTION AND WATERPUMPING

Contract number	: ESC-P-054-B(D)
Duration	: March I, I98I - June 30 , I983
Site of the installation	: Hoboken near Antwerp, Belgium
Power	: Rated peak power of array : 30 kW
Total budget	: 620.625 ECU(CEC contribution : I50.480 ECU, i.e. 24%)
Sponsors beside the CEC	: Programmatie van het Wetenschapsbeleid Nationaal R-D Programma Energie
Head of project	: P. Van Hal
Address	: E.N.I. N.V. Kontichsesteenweg 25 B-2630 Aartselaar
Members of Consortium	: GENSUN - C 50 - E.N.I. n.v. - Electrische Nijverheids - Installaties - G.M.B. n.v. - Generale Maatschappij van België - M.H.O. n.v. - Métallurgie Hoboken Overpelt - TRACTIONEL S.A.

Summary

The aim of this work is, in first instance, to achieve hydrogen gas-production and storage by means of solar energy, thus providing through a photovoltaic powerplant an autonomous way to accumulate energy.

The proposed project would also help to gain experience in implanting and connecting a photovoltaic array in difficult conditions, indeed the array is installed on a factory roof(avoiding floor space occupation) and in northern misty and industrial atmosphere.

Main system components are :

- The silicon monocristal photovoltaic cells were produced by Fabricable, following the ESAT-procedure(Prof. van Overstraeten) and incapsulated by IDE, before testing them at ISPRA.
  Each of 9I2 modules composed of 38 cells (dia. IOO mm) provides a nominal 33 Watts peak.
- A galvanised steel framework was drawn for easy adaption to the existing warehouse, with a approximative surface of 500 square meters. The tilt angle of the array was chosen at 30 degrees.
- Under the array, inside the warehouse, a control room has been built containing the main switchboard with the load management system,two centrifugal pumps, each rated at 7,5 kW and the data monitoring desk.
- The hydrogen production unit, with a capacity up to I2 kW, stands in the open air near the warehouse.
  The high performance electrolyser(alkaline water electrolysis) is a prototype, issued from a collaboration between SCK Research Centrum in Mol and E.N.I.

COMMISSION OF THE EUROPEAN COMMUNITIES

30 KWp PHOTOVOLTAIC PILOT PROJECT
HOBOKEN/BELGIUM

CONTRACTOR: GENSUN C-50
(E.N.I. - G.M.B. - TRACTIONEL - MH0)

ELECTROLYSER: SCK - MOL

I. Introduction

I.a/ Background and origin of the project

Contemplating the growing interest for new energy sources, four Belgian companies decided in 1980 to join their efforts to acquire the necessary know-how in the solar energy field.

The consortium shows as follows :

- Generale Maatschappij van België, a holding company, acting as catalysor between underneath mentioned companies.
- The engineering company Tractionel
- Métallurgie Hoboken Overpelt, active in the refinement of non ferrous metals(also offering the necessary space for the future erection of the GENSUN - C 50 project)
- and Electrische Nijverheids - Installaties , a contractor for electrical and instrumentation works (becoming the leader of the GENSUN - C 50 project).

E.N.I. was already involved in photovoltaic research work with Louvain-la-Neuve University, and studying an advanced water electrolyser in collaboration with the S C K Research Center in Mol.

A major problem encountered with solar energy systems concerns the storage of the production.
The consortium was prepared to investigate if the hydrogen vector could answer this question, and contemplated the direct coupling of a photovoltaic array to a water electrolyser. The power plant works completely autonomeous and supplies hydrogen gas which shall be stored in tanks.

Decision was taken by the consortium to present such a plant at the "CEC-R&D Photovoltaic Solar Energy" enquiry for pilot projects.

I.b/ Scope and objectives

The scope is defined as follows :

- 30 kW photovoltaic array
- directly coupled to a 10 kW water electrolyser
- and two DC driven waterpumps, each 7,5 kW rated
- through especially designed load management system, ensuring automatic working with priority users condition.

The objectives put forward are :

- Demonstration of reliable hydrogen production from a photovoltaic array
- Demonstration of direct coupling between a P.V. array and DC-motor
- Demonstration of possibility to store hydrogen with self sufficient plant
- Demonstration of installations of P.V. array on existing roof
- Investigation on P.V. energy production in northern and misty country
- Investigation on corrosion resistance in industrial atmosphere

II. The Project

II.a/ Generalities

Hereunder , condensed technical information is provided regarding the different components of the plant :

1. General description

The installation can be subdivided in 3 main parts(see appendix I) :

1/ The P.V. array, rated 30 kWp, installed on the roof of an existing warehouse.

2/ Under the array, inside the warehouse, a small control room has been built.
In this control room are installed the main switchboard with the load ma-

nagement system , two centrifugal pumps, each rated 7,5 kW, and the data monitoring desk.

3/ The hydrogen production unit, with a capacity up to I2 kW, stands in the open air, at a distance of about 50 m south of the warehouse. The unit is treated as an independent load with its own auxiliary control and alarm functions. Therefore there is also a separate electrical cubicle inside the hydrogen plant.
The system is designed for a service voltage of IIO V DC.

The steady control voltage of 24 V DC is assured by small lead-acid batteries. When the array is connected, the batteries are charged through 3 plain DC/DC converters with an output power of 750 W each.

The load management is achieved by a Texas Instruments PM 550 programmable controller, according to the logigram represented in appendix II. A 300 W, 25 V DC to 220 V AC commercial inverter, is used for the power supply of this device.

So the complete installation is self-standing, powered by solar energy. There is a possibility, but only for test purposes, to switch over to the grid, and the main switchboard contains an adjustable AC/DC supply.

Only the data monitoring system is supplied by the grid, since it was considered as a means to collect information only, without vital function.

2. Array support structure :

The array had to be installed on the south exposed side of an at 2I° inclined roof. The roof covering of corrugated asbestos plates, was not suitable for supporting any structure. This problem was solved by bolting and independent galvanized steel structure to the columns of the building. At the same time the tilt angle was improved to 30° with the horizontal. The south-side of the roof measures approximately 60 m by 8 m and is entirely covered by the array.

The access to the array is facilitated by a gangway along the edge of the roof and also by the cable ladders between the subarrays.

A subarray is composed by 8 horizontal groups of 6 modules. The I52 groups were assembled and wired in the factory on galvanized steel frames, which fitted perfectly between the trusses of the supporting structure.

The 20 trusses are connected with the building earthing.

3. Modules :

The 9I2 modules are made by I.D.E.-Fabricable. A module consists of 38 series connected monocrystalline silicon cells. The terminals and the centre branch are wired to a watertight box which contains two by-pass diodes, one on each group of I9 cells. The box is equipped with a bushing connector. The cells are encapsulated in EVA between two tempered glass plates. The frame in anodized aluminium is specially shaped in view of the fitting to the support structure. Rubber shock absorbers are fitted between the galvanised steel structure and the alu module frames in order to minimize possible deformation effect in the modules.

Modules characteristics :

- cell diameter : IOO mm
- nominal peak power : 33 W
- nominal open circuit Voltage : 2I V
- nominal short circuit current: 2,I A
- dimensions : IO78 x 440 x 50 mm
- weight : II kg

4. P.V. Array and cabling :

The array is composed by I9 identical sub-arrays. Each sub-array consists of 8 series connected groups of 6 parallel connected modules. In order to minimize the mismatch losses, the parallel connected modules were selected on a base of the same voltage output near the maximum power point, and the sub-arrays were formed by selecting groups with the most similar current outputs. So each of the 9I2 modules has its own, specific location in the array.

AMP twin lead SOLARLOK connectors are used for the parallel connection between the modules. The wiring is realized with VOB(Belgian Standard) 4 mm2 stranded wire, which is the extreme possible for the Solarlok connector. Wiring losses are minimized by connecting the module-groups at both sides, in a closed loop, as shown in appendix III.

A LED is installed across the output of each module-group, in the nearest junction box. Therefore the junction boxes have transparent covers. The LED's are an aid in the detection of a faulty module.

The sub-arrays are cabled with IO mm2 VOB wires to a terminal cubicle on the roof. Inside the cubicle, there is a three-position switch on each of the I9 sub-arrays, allowing the choice between the open circuit, short circuit, or connection to the main switchboard.

5. Main Switchboard :

The switchboard was built in the E.N.I. factory. It contains the contactors for the connection of the loads, the control and alarm relays and the interfaces between the analogical process parameters and the load management system. A simplified diagram is represented in appendix III.

Except the motorfield regulators, all the components are standard available parts. The control voltage is 24 V DC, supplied by a 75 Ah lead acid battery, assuring a night and day watching of the start up and alarm conditions. In case this battery should run low, a second 60 Ah battery, installed in the hydrogen switchboard, will take over the alarm signalisation.

Inside the switchboard a I4 kW adjustable rectifier, connected to the grid, makes it possible to test one of the loads, any time the manual mode is established. In the normal , automatic mode, the array-plane pyranometer decides in a first stage the closing of the array contactor feeding the DC/DC convertor, and in a second stage, the connection between the battery and the control circuits. From then on, the load management system is powered and will take over the process control. The first priority will be allocated to the electrolyser up to I2 kW. Above this power, the two 7,5 kW centrifugal pumps will subsequentually be switched on. Their speed will be regulated through PID loops.

Measurement instruments mounted on the front doors of the switchboard, show the values of the available solar energy, the system voltage, the control voltage, and the current intensities of the different loads.

6. Waterpumps

Two centrifugal pumps were put at our disposal by the firm ALLWEILER. The pumps are driven by 7,5 kW, IIO V DC motors, with independent excitation in 24 V DC. The speed is adjustable , up to 3200 r.p.m. The pump group includes also two non return valves, a pilot valve actuated by a 24 DC motor, a flowmeter, a level detector and a differential pressure switch. The pump characteristics are given in appendix IVa and IVb.

The water is sucked up in an existing pipe line coming from the industrial water epuration station of the M.H.O.-Factory, and is pumped into another pipe line which is connected to a water tower. The pressure head is 25 m à 35 m water column.

The level detector assures that the pumps cannot run without water, while the differential pressure switch operates the pilot valve.

When connected to the P.V. array, the speed control is carried out by P.I.D. loops of the load management system. When the installation is on the manual test mode, the speed control is achieved by potentiometers on the main switchboard.

7. Hydrogen Plant :

Even at low irradiation levels, the array can be connected to the electrolyser, without any kind of power-conditioning. The electrolyser is a prototype , worked out in collaboration between S.C.K. and E.N.I. , and is based on a new concept in advanced alkaline water electrolysis. (cfr.ref.nr.1)

An inorganic ion-exchanger membrane, based on polyantimonic acid(PAM) has shown its chemical stability up to 10.000 hours in NaOH solutions at 120° C, and its ion conducting and gas separating properties under electrolysis for 5.000 hours up to now(cfr.ref.nr.2)

Electrocatalysts based on non-noble metals, were deposited onto the perforated nickel plate electrodes. The electrolyser consists of 58 electrolysis cells in series, with a membrane area of 300 cm2.

The load management system will limit the current density to 0,4 A/cm2.

In this project, the produced oxygen is lost into the atmosphere. The hydrogen produced under a pressure of 0,11 MPa, is lead to a 500 l buffertank. The hydrogen gas is passed through a gas-dryer, and then compressed in the 1600 l storage tanks. When the pressure rises up to 6 bars, a signal is given to the M.H.O. service, which is responsible for the tank replacement.

The curve of the maximum power points of the array meets almost perfectly the electrolyser charasteristics. Future experience will tell us the relationship between the number of series connected electrolysis cells, the array output voltage and the hydrogen production.

8. Data Monitoring system

The data monitoring is a system totally separated from the rest of the installation and is powered by the grid. The hardware is composed by Thomson microprocessor, digital and analog cards. Every hour, the data are recorded on an ECMA 46 magnetic tape cassette, in the format required by the C.E.C. In order to obtain instantaneous information the system is completed by a Texas Instruments T.I. 840 Printer. Appendix V shows the bloc diagram of the system.

The recorded data are :

- Global horizontal insolation
- Insolation in the tilt angle of the array
- Max. wind speed
- Average wind direction
- Average ambiant temperature
- Cell junction temperature
- Array output energy
- Energy to each load
- Exploitation results
- Switching events.

Following measurement instruments and sensors are used :

- I pyranometer "Kipp & Zonen" CM10 (horizontal)
- I pyranometer "Haenni" Solar II8 (30°)
- I wind station "Lambrecht" I464G3RH
- A pt IOO sensor (ambiant temperature)
- 8 Ni-Cr-Ni sensors, sticked to the back of 4 modules in the array
- 2 waterflowmeters Aqua-metro
- Shunts and transducers.

Furthermore, the data monitoring system is used for the watching of the sub-array outputs. Taking into account the normal differences between the array outputs, any divergence will be recorded as an event.

9. Lightning protection :

Considering the low keraunic level of the region, direct lightning flashes to earth are very unlikely, and no protective devices were installed.

A direct stroke was never recorded at the M.H.O. factory, and furthermore, several high chimneys, topped by classical masts, surround the P.V.-array. Protection against induced transients was achieved by "TRANZORB" voltage suppressors. Two of these protections already burned out, probably because of overvoltage generated in the system. An alternative device shall be selected, after further investigation of the phenomenon.

10. Architectural and environmental aspects :

The P.V. plant is situated inside the M.H.O. Factory at Hoboken, about I5 km south of Antwerp, along the right bank of the Scheldt river. It is an industrial and dusty site, not visible from the public way. From an architectural point of view, one can rather say that the array has an embellishing effect on the environment. On the other hand, it is to be feared that the dust and the fall-out of the surrounding chimneys will have a harmful and even destructive effect on the installation.

II.b/ Testing apparatus and measuring equipment :

The data monitoring system is the principal tool for the evaluation of the efficiency of the installation. The data are collected with a sampling time of I min. Results are registered every hour on ECMA 46 tape, and can also be printed out. The inputs of the system are listed in § 8 of the previous chapter. Switching and alarm events are also recorded and at midnight the performances of the day are totalized and registered. Watching and comparing the sub-array output currents, the data monitoring system is also used as a first step in the detection of faulty modules. The next step is facilitated by the I52 LED's installed in the junction boxes on the roof, accross each group of 6 modules.

Another possible but not yet tested means to locate a faulty module is the use of a portable digital thermometer.

The total array output current, the array voltage, the insolation level, the control voltage and the currents to each load, are displayed by indicating instruments on the main switchboard.

Process variables and setting points can be read on the loop access module of the load management system.

The hydrogen production can approximately be stated by the number of I600 l tanks, filled up at 6 bars. The data monitoring system takes into account the quantity of water which is dissociated into hydrogen and oxygen-gas.

There is also a gas-flowmeter, allowing an accurate but not recorded measurement of the hydrogen production.

For safety reasons, a permanent gas-analysis device is installed in the oxygen outlet of the electrolyser. At a 2% hydrogen presence in the oxygen circuit, the electrolyser is cut off.

The waterpumps are also equipped with a flowmeter, which gives an impulse to the data monitoring system at each 100 liter revolution.

II.c/ Results :

1. Array

The first results to be mentioned, are the module characteristics as measured by the manufacturer. The complete array population is summarized in the following table, showing the average and the extreme limits of five typical values :

	Average	Min	Max
Short circuit current	2,24 A	1,67 A	2,49 A
Open circuit voltage	22,73 V	21,93 V	23,38 V
Fill factor	0,68	0,57	0,76
Maximum power	34,7 W	26,2 W	39,9 W
Voltage at max.power	17,62 V	15,94 V	18,72 V

The graph in appendix VI, shows the staggering of the maximum module power outputs, over the entire array population.

The CEC acceptance tests and other contradictory measurements on some modules, seem to demonstrate that the values announced by the manufacturer are noticeably too high for the low rated modules but are realy overestimated in the higher ranges. Whatever may be, the acceptance tests carried out by ESTI, show that the net output of the array is at least 27.300 W at standard test conditions, so that the contractual minimum is reached.

This means that the contractual rated peak power of 30 pk has been delivered.

2. Electrolyser :

The second result and principal aim of this project, is also fulfilled : the construction of a new type of high performant water electrolyser, and the coupling of this appliance with a P.V.-array.

Please find enclosed (appendix VII) the summary of measured figures when the electrolyser is connected to the P.V.-array. Appendix VIII shows the working points of the electrolysis, together with the array characteristics.

3. Waterpumps :

We like to state here that we realized the direct coupling of the P.V.-array to the DC pump motors, without any inverter and /or other conversion.The principle which was selected is as follows : an independent excitation ensures optimal starting couple and allows a progressive speed control between no load (1900 rpm) and full load (3200 rpm) conditions.

The field regulation electronic device was especially developped by E.N.I. for this application.

The main pump characteristics were controlled in situ, connected as well to the grid as to the P.V.-array, and recorded results are given in appendix IX.

II.d/ Analysis of results and comments :

1. Array :

1.1. Mismatch

Despite the large dispersion of the module characteristics, as mentioned in chapter 2c.a/, we succeeded to minimalize the mismatch losses at less than I,5 % of the array output. To achieve this result, various investigations by means of computer programs, were performed regarding Vmax and maximum power output on the module population.

Best results are obtained by operating as follows :

1/ in the first stage, grouping 6 parallel connected modules with similar Vmax characteristics ;
2/ Next step consisting in assembling 8 serial connected groups of 6 modules presenting similar current characteristics.
3/ Finally connecting the I9 sub-arrays in parallel.

1.2. Diodes and wiring losses :

It appears from our calculations that these losses amount approximately to I,6 %.

2. Dust :

One of the objectives of this project is to evaluate the variable performance of a P.V.-array in industrial and dusty environment.

During acceptance test, it was possible to verify that a natural dust deposit can cause a loss of performance of 4 to 5 % within a period of three days.

On the other hand, it was noticed that heavy rainfall had a cleaning effect to some extent. We assume that in this respect a certain stabilisation will occur.

3. Smooth coupling between P.V.-array and users :

An open circuit voltage of I7O V was measured on the array in isolated condition. Our main users (electrolyser and waterpumps) were designed for working around IIO V. Even when connecting manually at once the P.V.-array to above mentioned users, it was noticed that the working balance point was met immediately.A transient tension drop was observed when switching on the full tension to the DC pumpmotors.

III. Conclusions :

The main objectives which were set up and achieved, must briefly be repeated here :

- successful completion of the installation and operation of the various components, powered by solar energy
- hydrogen production with new electrolyser type
- D.C. motor drive by direct coupling to P.V.-array
- Development of a load management system, ensuring automatic power use following specified priorities
- Study of new complex system and collection of related know-how.

It is obvious that much longer experience in the operation of such an intricated system will be necessary in order to ascertain further valuable conclusions.

The future will enable to distinguish useful and fruitful information when said installation will be run under influence of numerous parameters and when sufficient records can be analysed.

IV. List of references :

Ref. 1 : H. Vandenborre, R. Leysen, H. Nackaerts and Ph. Van Asbroeck : A survey of a five year intensive R & D work in Belgium on advanced alkaline water electrolysis - Proceedings of the 4th World Hydrogen Energy Conference, California , U.S.A. , I3 - I7 June I982 - page II6

Ref. 2 : H. Vandenborre, R. Leysen and L.H. Baetsle : Alkaline inorganic - membrane - electrolyte (IME) water electrolysis - Hydrogen Energy - Vol 5, page I65 - I7I.

V. Appendices :

I. General lay-out

II. Logigram

III. Array Cabling and Bloc Diagram

IV. Pump characteristics

V. Bloc Diagram - Data Monitoring System

VI. Array Population

VII. Hydrogen Production Table

VIII. Electrolyser Characteristics

IX. DC Motor Characteristics

10 kW WATER ELECTROLYSER

ELECTROLYSIS STACK

COMPRESSOR SET AND GAS DRYER

HYDROGEN FILLING STATION

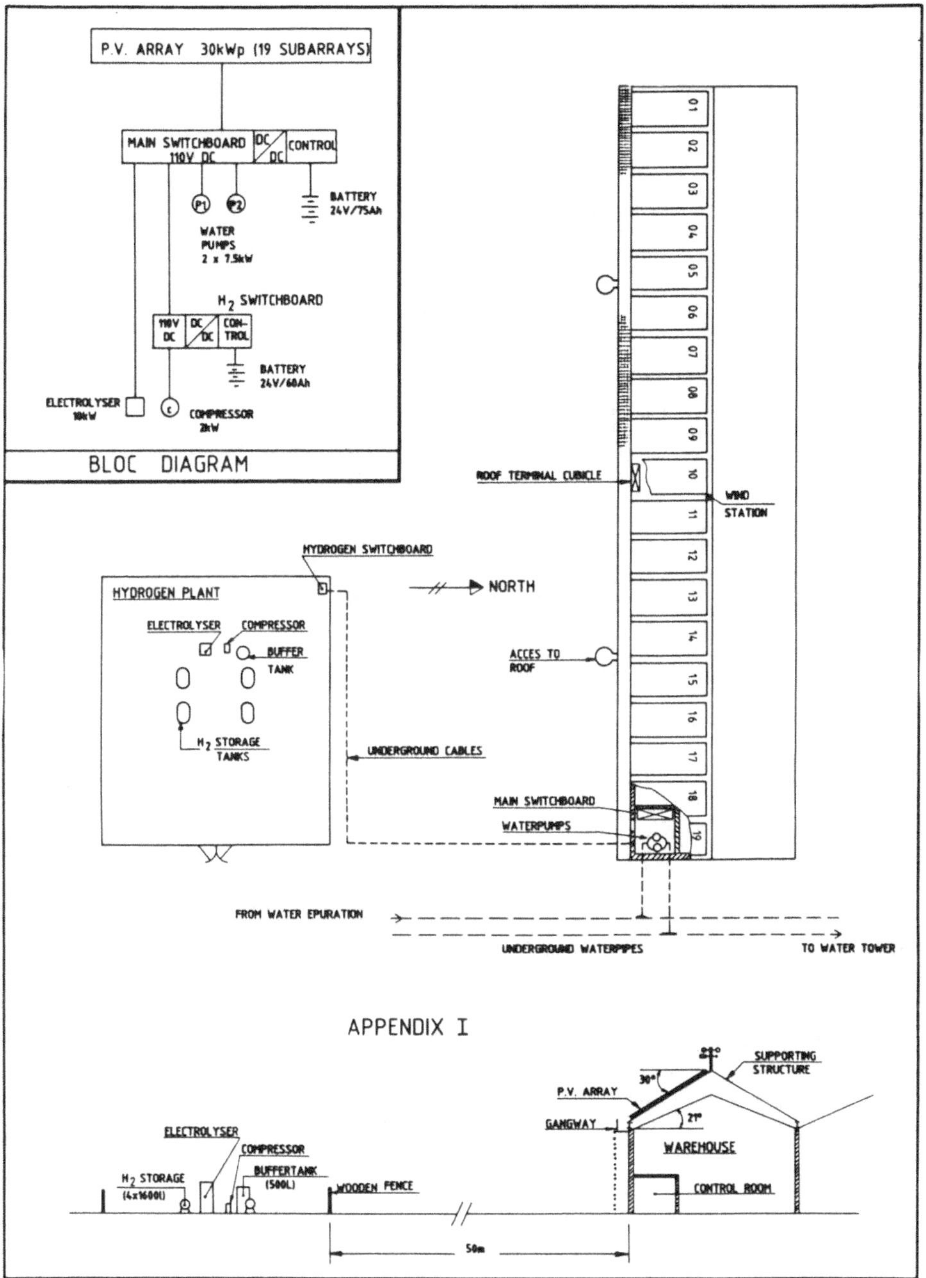

PHOTOVOLTAIC PROJECT GENSUN C50-General lay out

LOAD INCREASE

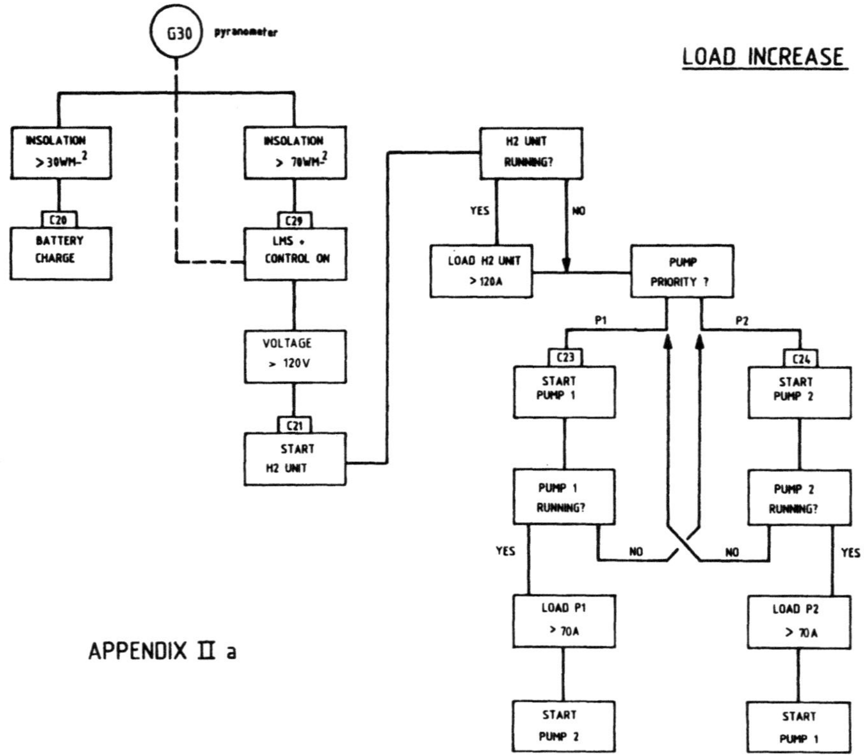

APPENDIX II a

E.N.I. 81250/201 p. 3/4

LOAD REGULATION

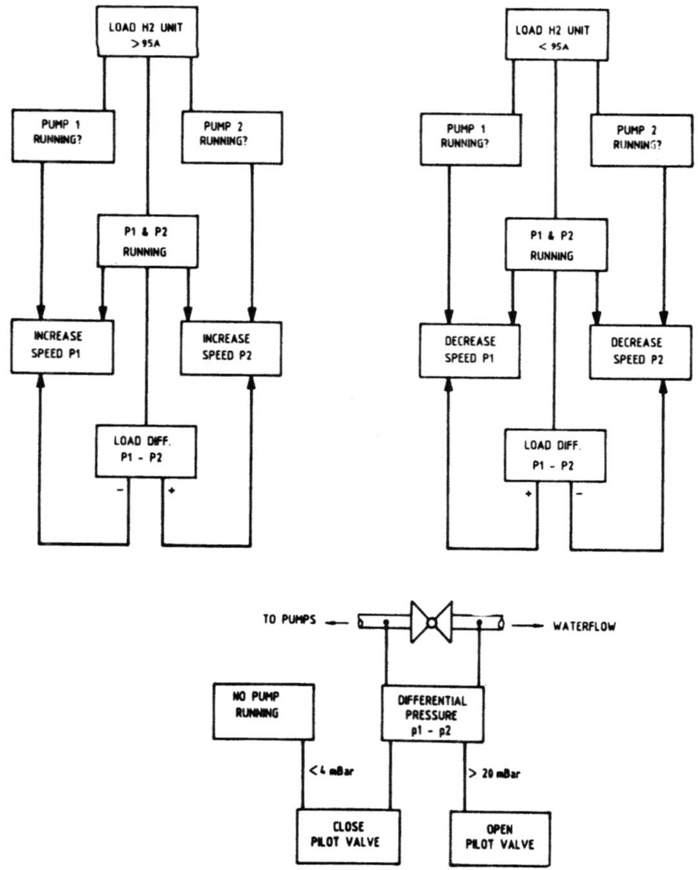

APPENDIX II b

E.N.I. 81250/201 p.4 / 4

LOAD DECREASE

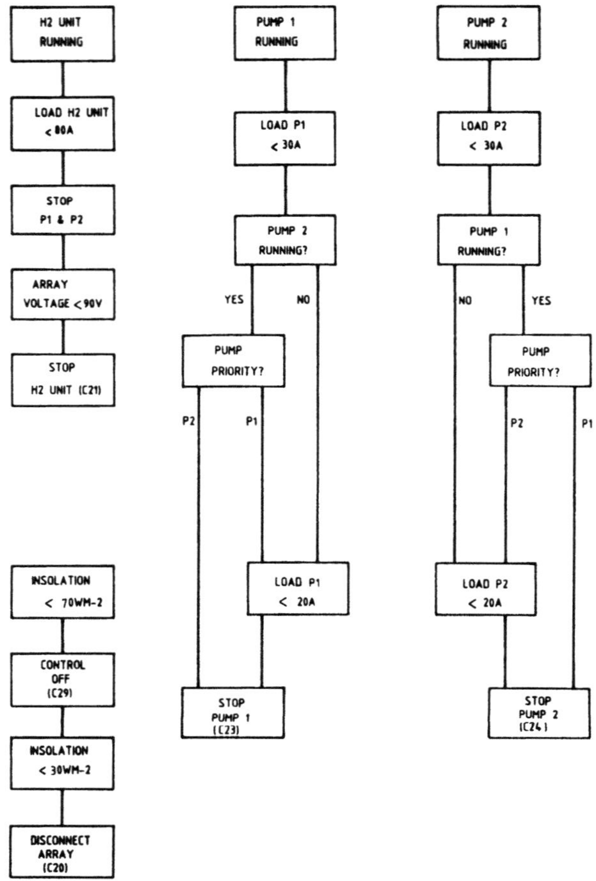

APPENDIX II c

E.N.I. 81250/201 p 2/4

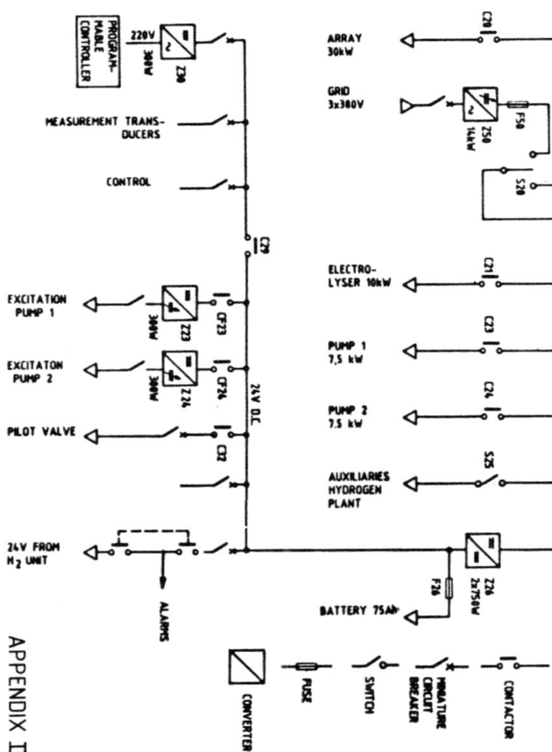

APPENDIX III

P.V. PILOT PROJECT GENSUN C50- Power distribution

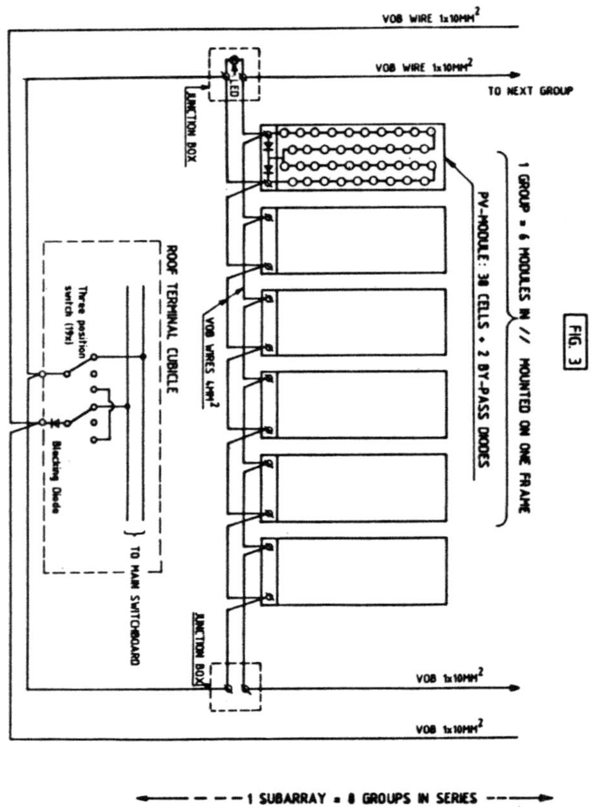

P.V. PROJECT GENSUN C50 - Principle of array-wiring

N.V. **E.N.I.** S.A. AARTSELAAR	**P.V. PROJECT «GENSUN C-50» Type NI-65-160 Allweiler pump characteristics**		1/2
			81250/012

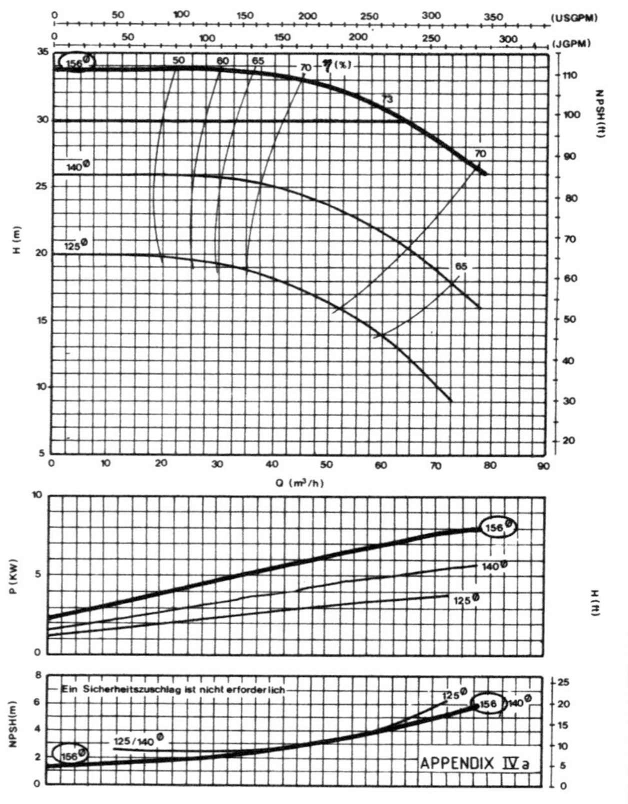

N.V. **E.N.I.** S.A. AARTSELAAR	**P.V. PROJECT "GENSUN C-50" Type NI-65-160 Allweiler pump characteristics**		2/2
		.	81250/ 012

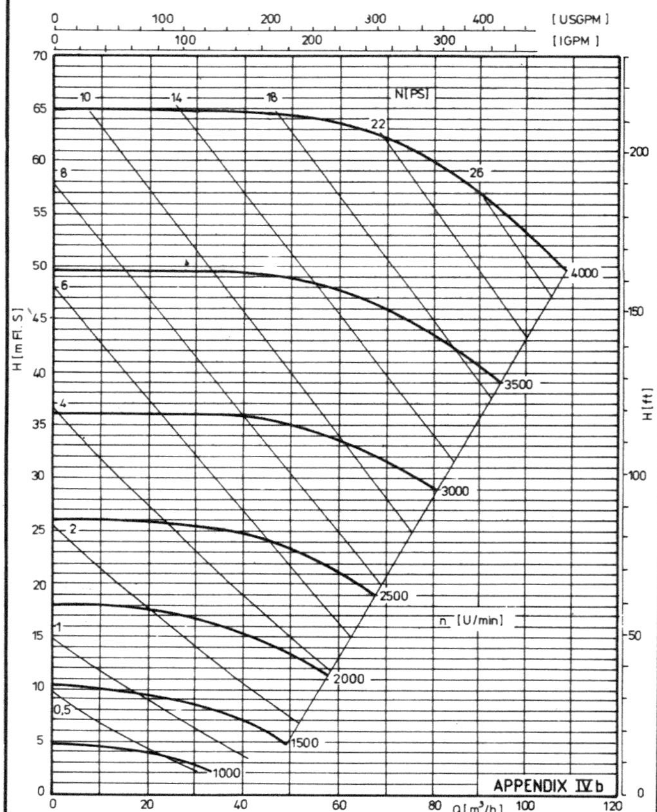

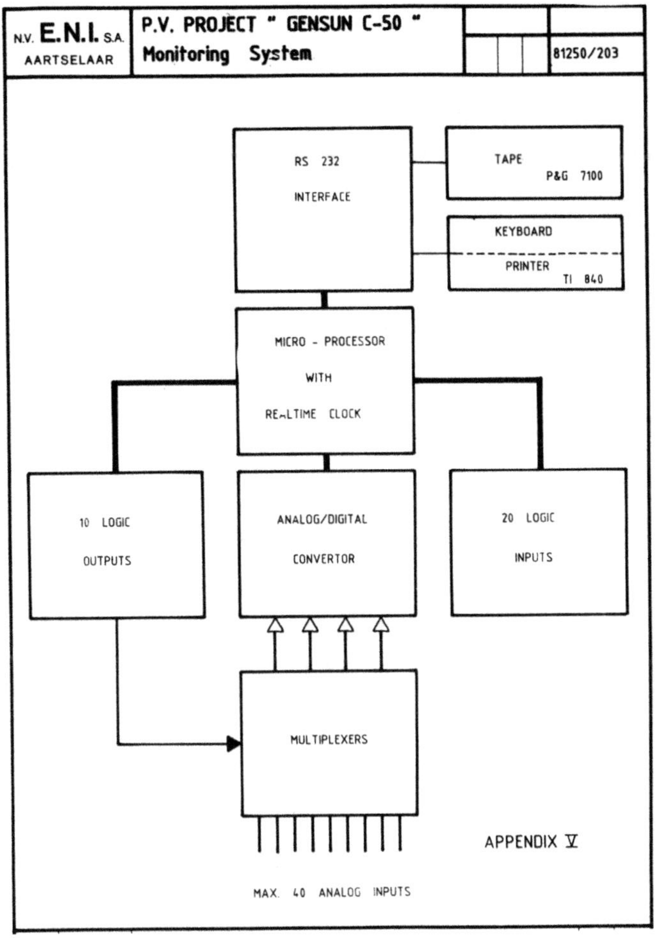

APPENDIX V

GENSUN C50 ARRAY POPULATION

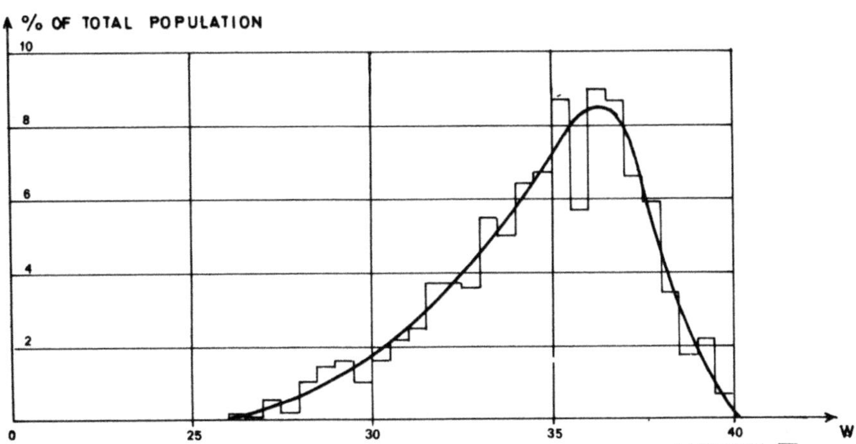

APPENDIX VI
MAX MODULE POWER

GENSUN - C 50 - PRODUCTION OF HYDROGEN (TEST I983, JULY I3th , AT HOBOKEN SITE) APPENDIX VII

SCK-ENI ELECTROLYSER COUPLED WITH PHOTOVOLTAIC ARRAY

Period	Temp.(x)	Current	Tens.(xx)	Power	Production(xxx)			Energy	Therm.Eff.
Minutes	° C	Amp.	Volts	kWatt	L/min	M3/period	M3/H	kWh/m3	%
O to 5	42	90	II5	II,7	36,3	O,I8I	2,I8	5,36	7O,8
5 to IO	5O	IOO	II7	II,7	4O,3	O,2OI	2,42	4,83	78,6
IO to I5	52	I25	II7	I4,62	5O,3	O,25I	3,O2	4,84	78,5
I5 to 2O	56	I2O	I2O	I3,2	49,33	O,246	2,96	4,45	84,5
2O to 25	58	IOO	IO8	IO,8	4O,3	O,2O3	2,42	4,46	84,4
25 to 3O	62	II5	IIO	I2,65	46,3	O,23I	2,78	4,55	83,5
3O to 35	64	I2O	IO7	I3,I4	48,3	O,24I	2,9	4,55	83,9
35 to 4O	66	IOO	IO7	IO,7	4O	O,2OO	2,4	4,45	85,4
4O to 45	67	IOO	IO8	II	4O,3	O,2OI	2,42	4,54	83,7
45 to 5O	68	II5	IIO	I2,65	46,3	O,23I	2,78	4,55	84,8
5O to 55	7O	I2O	IO7	I3,I4	48,3	O,24I	2,9	4,53	83,8
55 to 6O	7I	I25	IO7	I3,37	5O,3	O,25I	3,O2	4,42	85,9
Average over one hour		IIO,83	III,7	I2,38	44,6	O,223	2,68	4,6	82,6

(x) Max.temp. of electrolyte at outlet of electrolyser.

(xx)Net Voltage on electrolyser terminals

(xxx)Controlled with Ritter Gas Meter.

N.V. **E.N.I.** S.A. AARTSELAAR	**P.V. PROJECT « GENSUN C-50 »** P.V. module characteristics		81250/010

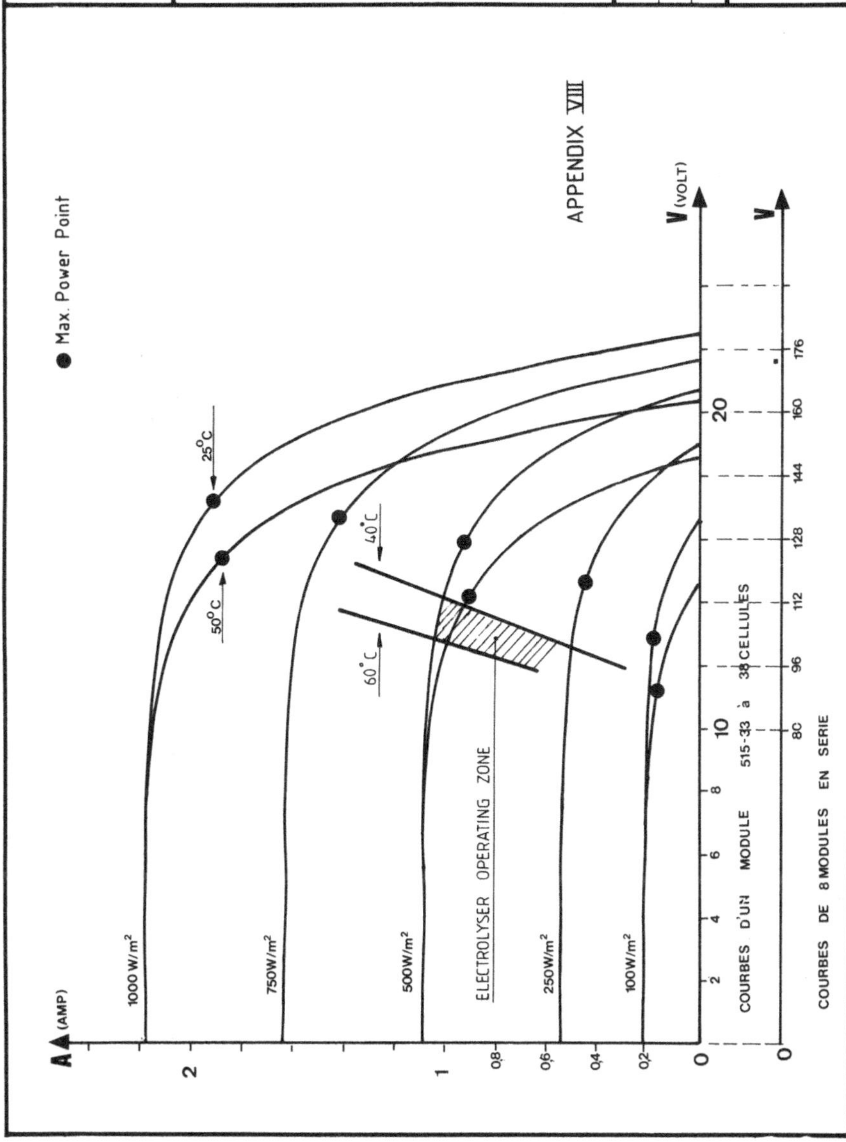

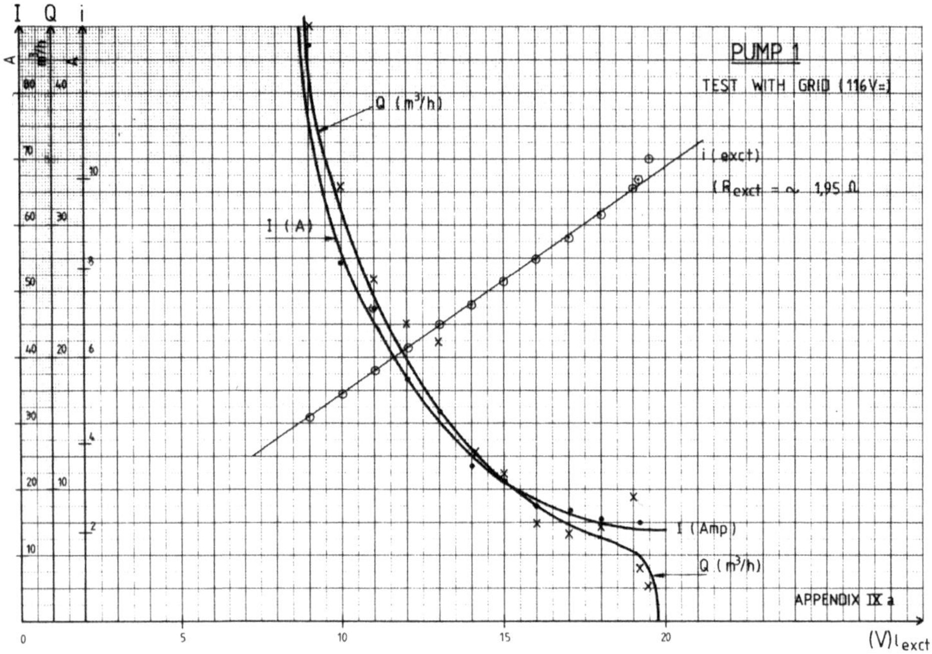

I Q i
A
m³/h
A
PUMP 1
TEST WITH GRID (116V=)
Q (m³/h)
i (exct)
(Rexct = ∼ 1,95 Ω
I (A)
I (Amp)
Q (m³/h)
APPENDIX IX a
80 40
70 10
60 30
8
50
40 20 6
30 4
20 10
2
10
0 5 10 15 20
(V) iexct

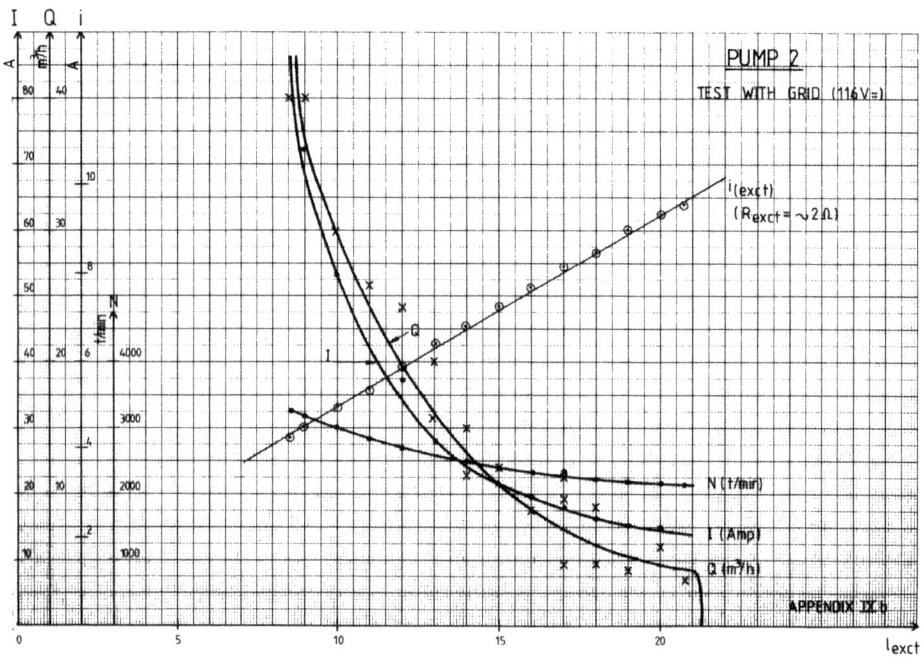

I Q i
A
m³/h
A
PUMP 2
TEST WITH GRID (116V=)
i(exct)
(Rexct = ∼2 Ω)
t/min N
Q
I
N (t/min)
I (Amp)
Q (m³/h)
APPENDIX IX b
80 40
70 10
60 30
8
50
40 20 6 4000
30 3000
4
20 10 2000
2
10 1000
0 5 10 15 20
iexct

MARCHWOOD PROJECT - 30 KILOWATTS

MARCHWOOD POWER STATION SITE

Contract number : ESC-P-069-81-UK(H)

Duration : 34 months 1st March 1981 - 31st December 1983

Site of the installation : Marchwood Power Station Site near Southampton, UK

Power : Peak PV power : 30kW, Voltage 415V, 3 Phase, 50Hz

Total budget : £900,000 (EEC Contribution : £205,000)

Sponsors beside the EEC : BRITISH PETROLEUM
UK GOVERNMENT

Project Leader : R D W Scott, C Eng MIEE

Address : BP Solar Systems Ltd
Windmill Road
Haddenham
Aylesbury
Bucks HP17 8JB
England

Summary

BP Solar is about to construct a solar generation plant with a peak output of 30kW on the CEGB Marchwood Power Station site. The power output is to be integrated into the Marchwood site distribution system either to supplement the power taken from the grid network or to support individual, isolated, selected loads. Experience gained from this demonstration project will assist in the design of future rural electrification and grid linked solar generators. The system design includes a number of novel concepts. The highly redundant modules are to be mounted on a newly designed low cost structure of an easily assembled design incorporating a simple interwiring system. The DC controls facilitate four modes of operation with any combination of battery connected, grid connected and autonomous generation. The inversion sub-system is designed to accommodate these modes at a high overall through-put efficiency. Finally the monitor will provide a comprehensive range of data for instantaneous display and subsequent data analysis.

1.0 INTRODUCTION

As part of their Solar Energy Research and Development Programme, the EEC is sponsoring selected photovoltaic power generation pilot projects in Europe. As participants in this programme, BP Solar has undertaken to assess the potential for solar power generation in the UK.

A 30kw solar power system is shortly to be constructed on the site of the CEGB Power Station at Marchwood, near Southampton, the power output from which is to be integrated into the existing electricity distribution system on the Marchwood site, either to feed into the grid network or to supply independent segregated loads.

2.0 PROJECT DEFINITION

BP Solar has designed, developed, and will be installing shortly, a solar power generation plant with a peak power rating of 30kW on the site of the old coal yard at CEGB Marchwood Power Station. The output from this plant will be directly coupled into the existing Marchwood site distribution network at 415V, 3 Phase, 4 wire, 50Hz, or arranged to supply segregated loads.

The solar power generation plant consists of a solar array comprising 960 modules, power inversion equipment, a storage battery and associated DC controller and a comprehensive monitoring system.

The output from the solar array is connected to a 40KVA forced commutated power inverter which converts the DC output from the solar array into AC power suitable for synchronising with the Marchwood distribution network. A 400Ahr, 240V storage battery is connected in parallel with the array to store or supplement its power output.

A comprehensive monitoring system is to be installed which will measure and collect, via a data processing computer, operating parameters, environmental conditions and other associated information so that a detailed off-line analysis can be made to determine the validity of this form of power generation in supplementing conventional generation schemes and to assist in the development of future commercial installations of this type.

3.0 MODES OF OPERATION

The following modes of operation are to be considered during the commissioning trials and the design includes for the necessary flexibility for these modes to be operated individually according to an agreed time cycle:-

1) Array output synchronised to the grid network via the inverter without battery storage connected.
2) Array output connected to a selected load system uncoupled from the grid network - "standalone" mode - without battery storage connected.
3) As in 1) with battery storage connected.
4) As in 2) with battery storage connected.

A block diagram indicating the proposed arrangement is shown in Fig 1. Mode selection is via a switch on the DC control panel.

4.0 SITE

The initial designs for CEGB Marchwood Power Station allowed for the use of coal-fired boilers for steam raising, and as a result an extensive coal handling plant was installed. The coal was to be stored on a vast concreted site adjacent to the jetties on Southampton Water, convenient for off-loading the coal from barges.

In the event this concreted coal storage area was never used since during the commissioning of the power station the boilers were converted for oil-firing.

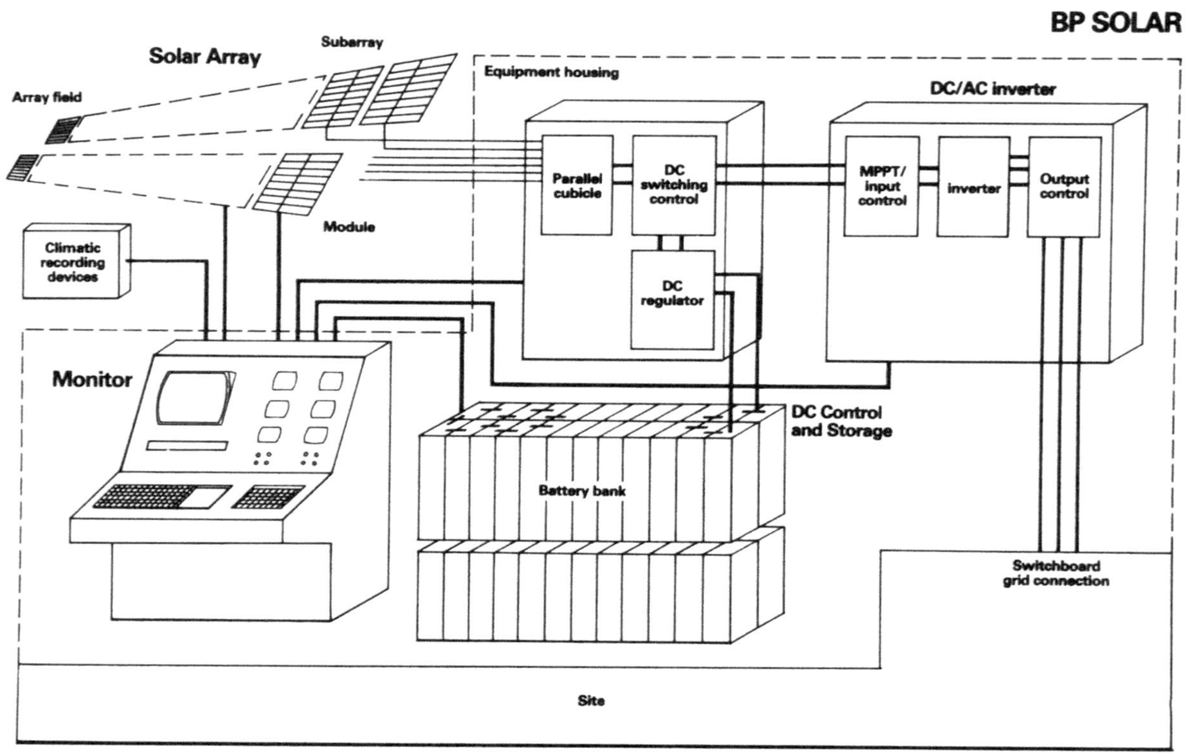

FIG 1 BLOCK DIAGRAM

Sections of the coal storage yard are now leased to private concerns under legal agreements for the installation of experimental projects. BP Solar have negotiated an agreement for the lease of an area of approximately 3000 square metres for the installation of the solar equipment associated with this project. The lease agreement which BP Solar has signed includes clauses which satisfy local authority environmental planning requirements and requirements stipulated by UK government departments.

The site is attractive from the point of view that little or no civil modifications are required and all equipment can be easily installed on site and bolted down to the existing concrete pad. The site installation period is therefore minimal.

It is possible that the leasing agreement will be terminated before the contract with the EEC expires, in which event arrangements will be made to transfer the installation to an alternative site, possibly one of the islands off the north coast of Scotland. The design of the installation will take this possible move into account.

5.0 CLIMATIC CONDITIONS

Marchwood is situated on the south coast of the United Kingdom and as such therefore encounters almost the most favourable climatic conditions of anywhere in the UK.

Extensive climatic records for this area are available and therefore PV array performance can be predicted accurately. In addition, array supporting structures have been designed to withstand peak wind velocities gusting to 150Kph.

Computer analysis of the available data has been used to calculate the anticipated output from the solar array at all times of the year. This will be used to select the fixed loads for "standalone" operations and also to compare with the monitored performance of the system.

6.0 LOAD CONNECTION

The output from the solar installation will be integrated into the existing Marchwood site distribution network via an existing 415V, 3 Phase, 50Hz switchboard currently supplying a variety of single and 3 Phase loads. Typically the loading on this switchboard ranges between 10kW and 30kW.

In the grid connected modes, the output from the solar installation will supplement the power supplied from the grid network and in cases where the solar output exceeds the load requirements for the switchboard, this excess power will flow into the grid network via existing 415/3300V circuit transformers.

7.0 SUB-SYSTEMS (refer Fig 1)

The installation is comprised of five basic sub-systems interconnected to produce the complete installation (see Fig 1).

1) Solar array and supporting structure
2) DC controls and battery storage
3) DC/AC inverter
4) Data monitoring system
5) Site, housings and interfaces with them.

7.1 Solar Array Field

The solar array field consists of 960 solar modules. Each module contains 36 individual solar cells of about 1 watt. These cells are encapsulated to form a rectangular solar module. The cells are laid up between layers of clear EVA and the laminated module is formed by interposing the cells between a toughened glass front sheet and a white water-resistant plastic backing sheet, which is composed of an aluminium foil layer sandwhiched between layers of white Tedlar. The laminate is sealed in a vacuum before the anodised aluminium edge frame is fitted.

The 36 cells are arranged in two strings each incorporating a by-pass diode. The positive and negative connections from the module are each terminated in a weatherproof plastic terminal box, which also incorporates the by-pass diodes.

Each module provides a peak output of approximately 33 watts at a nominal terminal voltage of 16 volts DC.

The modules are arranged in sections each containing 8 modules. The outputs from the modules in each section are connected in series. Two adjacent sections each of eight modules are connected in series to form a sub-array of 16 modules with a nominal terminal voltage of 240 volts DC (see Fig 3). The positive and negative output leads from the sub-array are terminated in a junction box incorporating an isolator.

The output from each of the 60 sub-arrays is transferred to the control cabin via a 2 core PVC insulated cable where the outputs are paralleled (see Fig 1).

The layout of the array field is shown in Fig 2. The spacing between rows has been maximised to limit shadowing effects within the constraints of the site boundary.

The design of the sub-array support structure is detailed in Fig 3 and Fig 4.

The sub-array is supported on a galvanised steel framework. The fixing feet, items 1 and 5 allow vertical adjustment to take up uneveness in the existing concrete foundation. Position of the front legs, item 3, provide for the lining up of adjacent sub-arrays so that each row appears architecturally pleasing. The front legs are clamped to the front fixing feet using a bracket, item 2. One front fixing position is used for two adjacent sub-array structures but additional bracing is incorporated every fourth sub-array, item 4. The modules are tied together mechanically using the bars item 7 and 9.

The sub-arrays can be inclined according to the season at three different angles. 55° +/- 15°.

The angle of inclination is simply achieved by choice of the appropriate bolt position on the lower angled support member, item 6. Adjustment of the angle of inclination is easily achieved by transfer from one bolt position to another. Three angles of inclination are available. The structural design also allows the sub-arrays to be lowered to a good working height for maintenance purposes (see Fig 4).

7.2 DC Control and Storage System

The DC system comprises:-

a) 400 Ahr storage battery
b) Paralleling panel
c) DC control panel

7.2.1 400 Ahr Storage Battery

The 400 Ahr storage battery is of the low loss lead acid type, specially developed for photovoltaic applications. It has been designed to optimise the characteristics for the charging regime typical of solar systems. The design therefore aims for high efficiency and low water loss without the poor cycle life and rapid aging typical of pure lead plated batteries.

The positive plates of the battery are of an alloy of lead with a small proportion of impurity, specially developed for strength and cycling ability. The plates are wrapped to extend their life with a sheet which

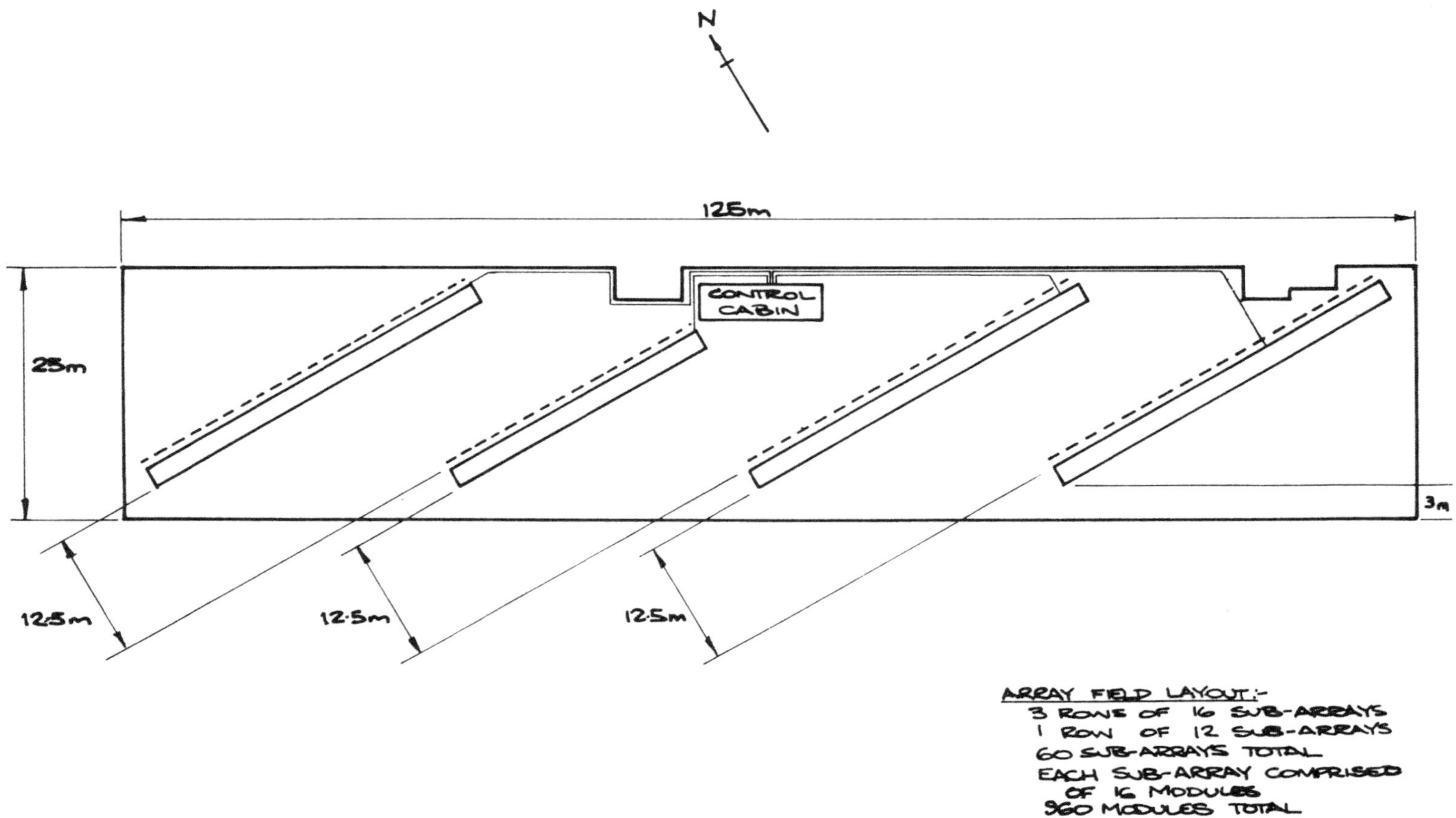

FIG 2: ARRAY FIELD LAYOUT

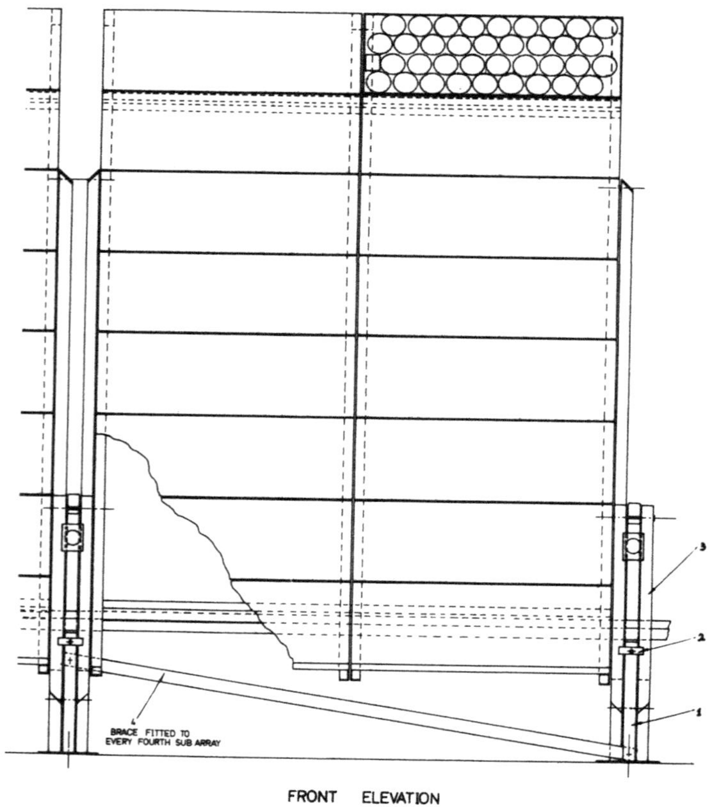

FIG 3: SUB ARRAY CONSTRUCTION

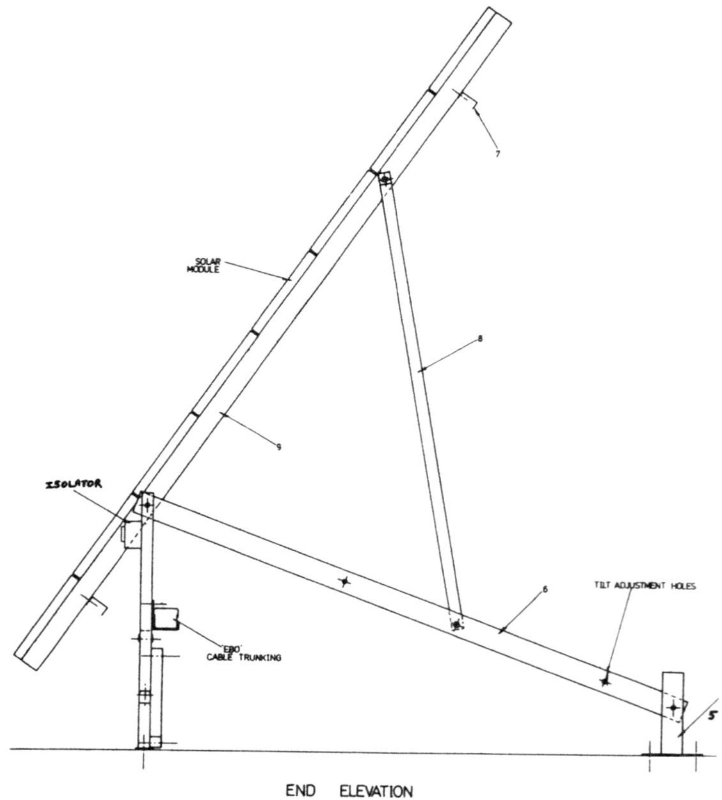

FIG 4: SUB ARRAY SUPPORT STRUCTURE

minimises stratification and packed and mounted in tough polypropylene cases. Sufficient electrolyte is provided to allow for long periods between replenishment.

All cells are nominal 2 volts units and are series connected to give a nominal terminal voltage of approximately 240 volts.

The battery cells will be mounted on a tiered epoxy coated metal stand and housed in a separate ventilated room in the control cabin.

7.2.2 Paralleling Panel

The paralleling panel parallels the individual strings on to positive and negative bus bars which are then taken to the DC control panel. The circuit diagram is shown in Fig 5. The output from each sub-array string is fed through a blocking diode and a selector switch, the switch allows the string to be connected to either the bus bars or a test load.

The test load facility is provided to enable an individual string characteristic to be checked by measuring the voltage and current as the load is varied from zero load to practically a short circuit.

Local monitoring on the panel comprises, test load voltage and current array voltage and current, and array power.

Each string is also fed through a transducer which provides a DC voltage signal proportional to the string current, the DC voltage signal being galvanic isolated from any other string. This signal is taken to terminal blocks to interface with the computer providing a monitoring facility for the individual strings. Similar DC voltage signals are also provided via transducer interface modules for the total array current and the array voltage.

An earth fault monitor is fitted to give indication both locally and via contacts to the computer should the array develop a leakage from either 'pole' to ground, the relay is adjustable for sensitivity between 100KΩ and a few ohms.

7.2.3 DC Control Panel

The DC control panel provides the switching mode selection logic for the operation of the system.

The input terminals for the array DC and the battery are provided at the top left front of the panel and the DC output for the PCU is situated at the top right front of the panel.

The array input actually enters from the side and the DC output also leaves through the side thereby keeping the power bus bars inside the suite and only the battery power leads enter directly through glands at the top.

Also situated at the top of the cubicle are the control interconnection terminal blocks for connecting to the PCU and the PMP and again the looms go through the top sides. The analogue transducer interface signals for the battery voltage and current etc are brought to terminals at the top centre front for connection through to the PCU cubicle. The 240V bias supply is also brought into this cubicle at the top centre.

The front door panel fixtures comprise the mode selection switch, system on/off switch, battery voltmeter, battery centre zero ammeter, battery isolator handle, array isolator handle, PCU input DC isolator handle and a mimic panel.

The mimic panel comprises a line diagram of the system showing isolators, contactors, main sections, etc and is complimented with LEDs to give the status information, also incorporated is the PCU fault signal computer read out meter.

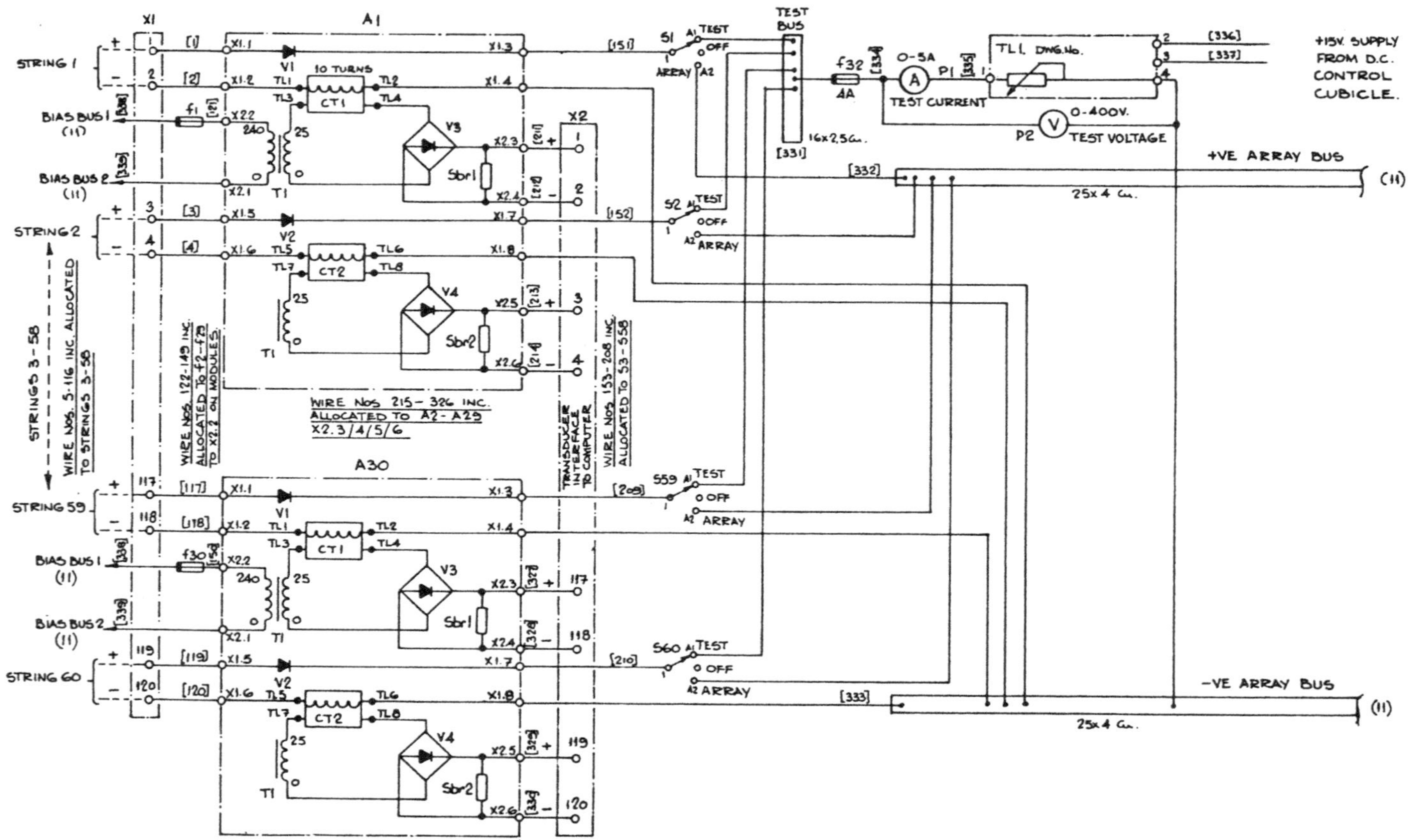

FIG 5: PMP CIRCUIT DIAGRAM

7.3 Power Conversion Unit

The power conversion unit changes the DC array power into AC and then either feeds this power into the grid or to a stand alone load dependent upon the mode selected. The main assemblies within the PCU comprise input filter, run up current limiter, impulse commutated thyristor bridges, output filter, control unit and maximum peak power controller. The circuit diagram is shown in Fig 6.

7.3.1 Power Stage

The method used to control the output voltage is that of step pulse technique; the two impulse commutated thyristor bridges are phase related and then fed on to two transformers which have the outputs added in a zig-zag configuration.

The advantage of this method being that harmonics are only generated at and above the 11th harmonic, and consequently are easier to filter, than would be required for traditional 3 Phase designs that only use zig-zag addition and not step pulse techniques.

The flow of power to the grid is achieved by connecting the PCU to the grid via inductors and controlling the phase relationship of the PCU with respect to the grid; if the PCU is leading in phase the power will flow from the PCU to the grid, the greater the lead the greater the power flow etc.

7.3.2 Control Stage

The main elements in the control unit are:-

- frequency generator
- thyristor gate switching amplifiers
- control logic for thyristor gate drive waveforms
- stop/start sequence logic
- regulator
- current limit amplifiers
- fault detection logic
- phase comparison of grid and PCU
- phase angle adjustment amplifiers
- AC and DC voltage monitors
- grid frequency monitor
- battery charge control regulator

The PCU control unit sets the correct sequence triggering of the thyristors on the impulse commutated bridge and accordingly sets the phase regulator to give the correct output voltage for the different input conditions and output demands. The frequency generator is synchronised with the grid frequency when the unit is connected in one of the modes of operation 'grid connected', otherwise the PCU will free run at its own natural frequency of 50Hz.

In the event of an output short circuit the electronics will react very quickly to ensure the output voltage is reduced by adjusting the phase relationship of the thyristor triggering and bridge phase relation such that the output current is held constant and no damage is done to the PCU inverter bridges etc.

7.3.3 Maximum Peak Power Controller

Because of the solar cells dependence upon environmental conditions, the system requires a means of adjusting the output load demand so that maximum benefit can be achieved from the power developed by the solar array. This function is performed by a microprocessor unit which monitors the voltage and the DC current at the input of the PCU. The DC voltage

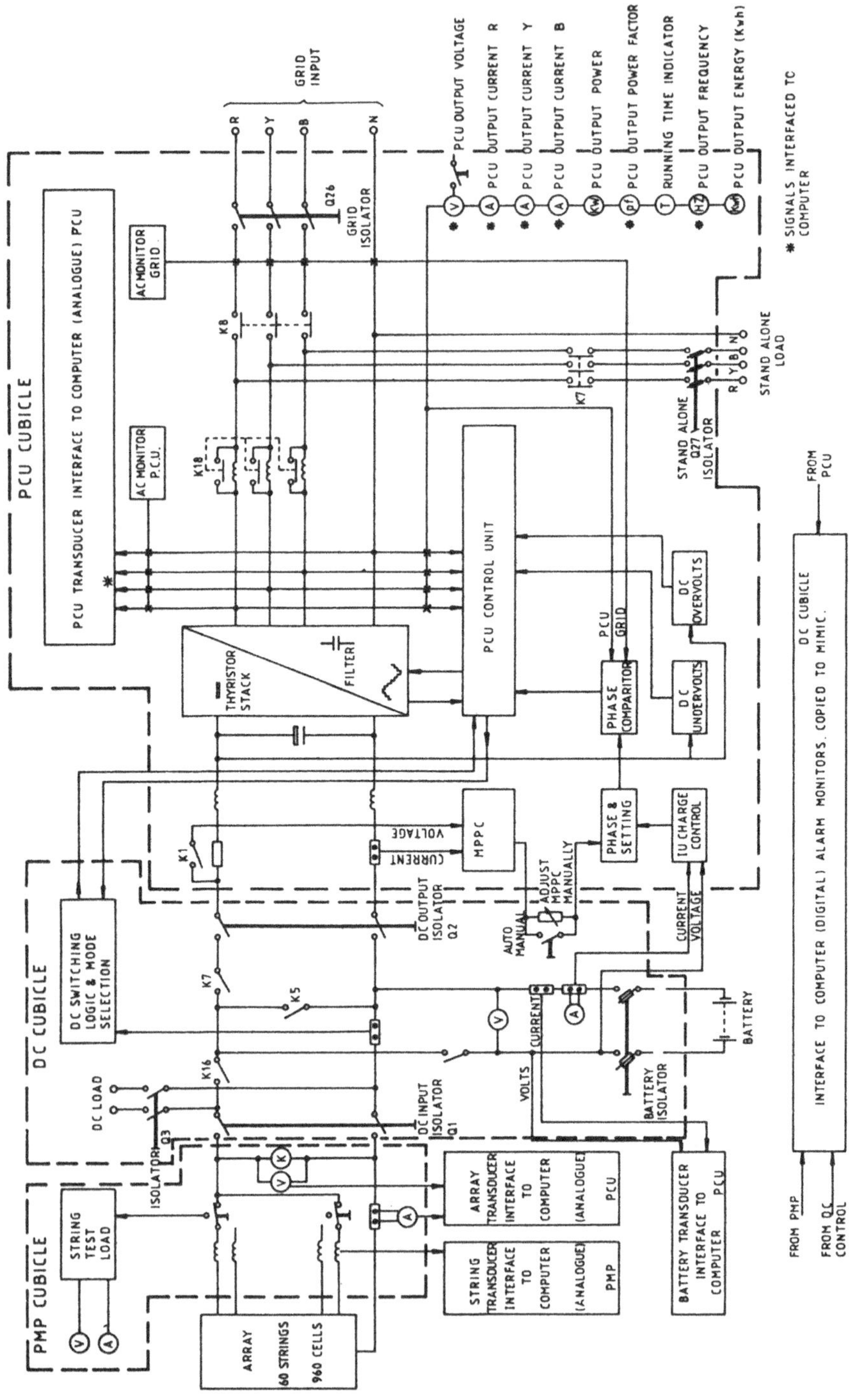

FIG 6: PCU CIRCUIT DIAGRAM

and current are multiplied together to give a signal representing power and its value stored in the µP memory, the PCU is adjusted and the input power measured again, compared with the previously stored value and the difference used to adjust the PCU to feed more or less power depending on the difference being positive or negative accordingly.

When the point of maximum power is reached the control swings either side of the finite actual value by approximately +/- 1%.

If the output of the solar array sinks to a value below the no-load losses of the PCU, then power flow from the grid is used to maintain the PCU no load losses for a period of approximately 2 minutes; this operation being to allow for clouds passing over, etc. After the 2 minutes, if the solar array has still insufficient power to maintain the PCU no load losses (ie going to night time) the PCU will switch off; the array will be shorted by a contactor in the DC control panel and the array current monitored, the PCU will only restart again when the monitored current reaches a value such that the no load losses could again be supported by the array; (this applies of course to the mode of operation "grid connected no storage" only).

The MPPC has also the facility to be controlled manually, the switch and potentiometer being located on the front door of the DC control panel. When the MPPC is in automatic and the array power has become sufficient to allow the PCU to switch on the MPPC samples the input power at a frequency of 0.5 to 1Hz and adjusts the PCU at this rate until the maximum peak power point is reached.

7.3.4 General Design Parameters

Each stage of the PCU has been specifically designed with optimum operating efficiency requirements in view whilst not sacrificing compatibility between stages. The following overall system efficiencies at various loadings are envisaged although these are considered conservative and further improvements may be achieved during the testing phase.

100% load - 92%	75% load - 90%
50% load - 88%	25% load - 70%

Stability and safety of operation have been carefully considered in the overall design of the PCU. Large, rapid swings in array field input as well as switching surges on the grid network can be accommodated without affecting the smooth operation of the PCU. The physical layout of equipment in the PCU cubicle will eliminate danger to personnel whilst still allowing easy access to all component parts.

The modular design of the PCU will allow automatic rearrangement of the various stages to facilitate operation in any selected mode as dictated by the master selector switch on the DC control panel.

A mimic diagram will be incorporated on the front of the PCU cubicle which will be so arranged to indicate the line diagram for the operating mode selected. In addition indicating instruments and banks of lamps to indicate alarm and fault conditions will be included adjacent to this mimic diagram.

7.4 Monitoring System

In designing the monitoring system for this plant a number of key parameters were identified. In view of the experimental nature of the plant the system had to be flexible and easily expandable in terms of plant inputs, computing power and data storage. Also, because of the proposed unattended operation of the plant and monitoring system for periods of up to 2 months over the five year period, the equipment had to be reliable and secure. The limited life of 5 years also required the system to be re-usable with or without reconfiguration of both hardware and software. Finally, because of the sales demonstration role which the plant was to fulfill the system facilities had to be comprehensive and attractive. To satisfy all of these demanding diverse requirements the monitoring system shown in Figure 7 was configured.

The various analogue and digital inputs from the photovoltaic cell arrays and invertor are marshalled into a self-contained plant interface cabinet. This is a Micro Consultants IRIS Level 1 system which has been programmed in a BASIC like language to provide the plant input scanning and first level conversion routines such as thermocouple linearisation. In this unit a number inputs are also averaged over one minute periods. After initial conversion appropriate inputs are checked against alarm limits and any violation notified to the main processor which is a Digital Equipment PDP 11/23 PLUS computer.

The PDP 11/23 computer provides all of the higher level facilities which include the data archiving, alarm annunciation, operator interface, graphic displays and report printer. For the routine unattended data archiving the larger 10 megabyte removable cartridge disks are used. The data files are both cyclic for the short term hourly, daily and monthly files, and linear for the long term yearly and five year files. It is possible to dump selected files of data to floppy disks for removal and subsequent processing on other computers. The system is configured for 3 different file dumps including one specifically formatted for data required for the EEC.

The most visible aspect of the system is a colour visual display unit on which can be displayed both alphanumeric reports and high resolution graphs and histograms. The facilities are available through a selection of operator commands which allow the operator to define the start and finish time of the data to be plotted and the time interval of the samples. All of the displays, both alphanumeric and graphical, can be hard copied to a colour matrix printer connected to the VDU. Two conventional printers are also provided, one for recording alarms and significant events and one for printing tabular reports. Alarms are also announced on the top line of the VDU and, because the control room is not normally manned, are also relayed to a lamp alarm in an adjacent gatehouse.

The unmanned operation of the system has also imposed the very important requirements for security of data and system recovery. In order to overcome the potential problems of re-start after a mains power supply failure a battery backed-up real-time clock is incorporated. After a power failure and automatic reload and restart of the system the data archiving task is able to identify the affected files and fill the appropriate records with invalid data markers. This invalid data marker is also used to highlight failed or suspect plant signals during normal scanning and if any of the subsequently calculated averages have insufficient valid results, as compared with a configurable limit, an alarm is generated.

Hence, by selecting proven and reliable hardware with a flexible and mature operating system such as RSX-11M and writing the application software in a structured language such as PASCAL, it has been possible to configure a monitoring system which will meet all the original design objectives. Sufficient flexibility and expansion capability has been incorporated to allow the addition of suitable facilities such as modelling, optimization and control. Although for this experimental application the system is constrained to data acquisition and display other opportunities exist on production systems. It would be possible to extend such a system to incorporate on-line modelling with optimization together with load shedding capability. In this way large unattended generating units

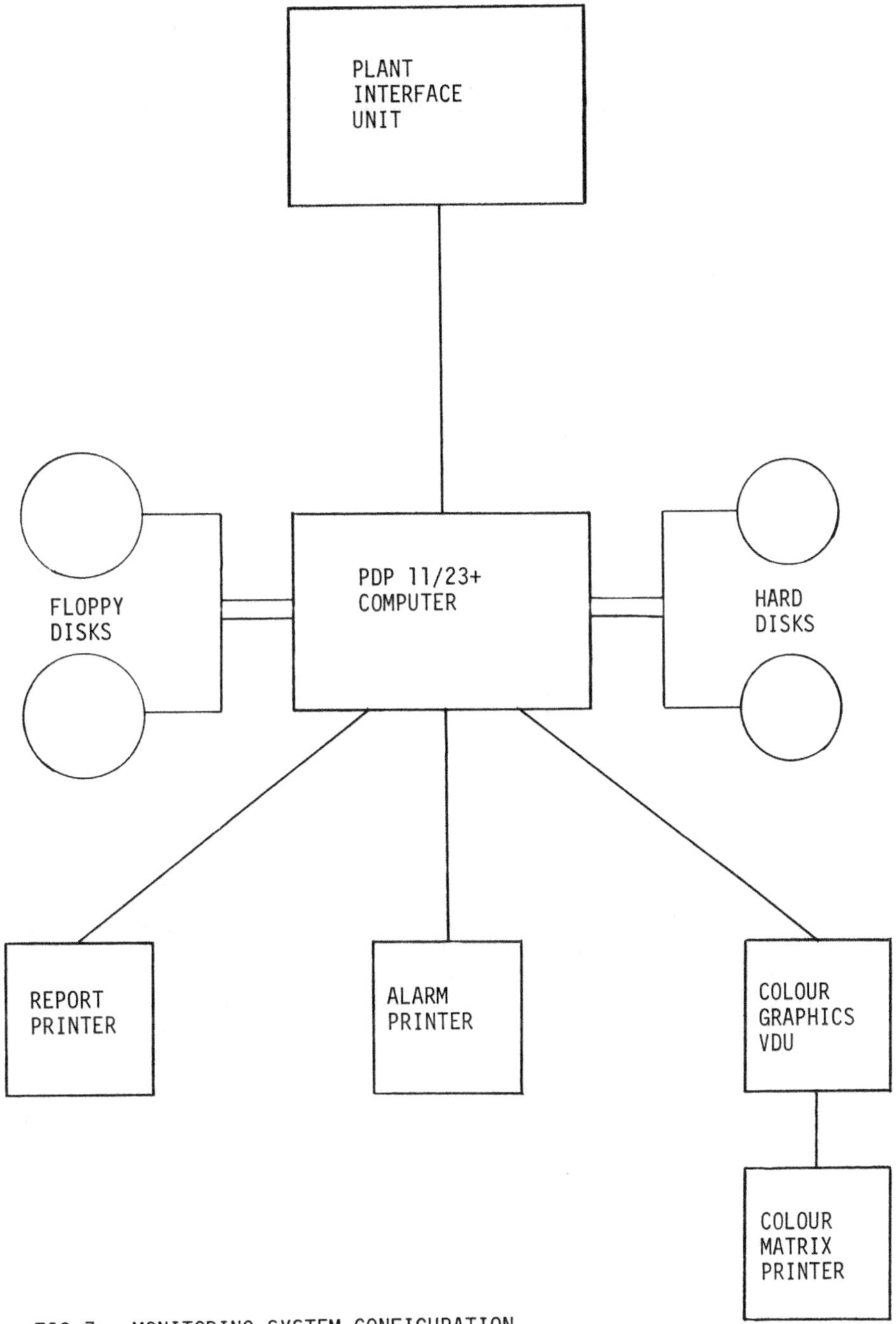

FIG 7: MONITORING SYSTEM CONFIGURATION

could be preprogrammed with various operating strategies.

7.5 Site, Housings and Interfaces with them

Installation of equipment at site will be arranged under a local site contract. The installation programme should run smoothly since little is required in the way of civil works. The site boundary will be marked by a post and chain link security fence with access gates. The outputs from the 60 sub-arrays will be cabled above ground to the control cabin. Cables for power and monitoring duty will be segregated and cleated on separate adjacent GRP enclosed cable trays mounted on the sub-array structures or fixed to above ground unistrut type supports adjacent to the control cabin.

The control cabin will be of the double skinned GRP design with integral metal skid base and lifting lugs. Three separate rooms will be provided. The battery room will house the battery cells mounted on a tiered epoxy coated metal stand and will include roof ventilators. The control room will house the DC control/paralleling cubicle and PCU equipment, and will be cooled. The data monitoring room will be fully air conditioned to house the data base/computer equipment and will be designed so that visitors to the installation can be pleasantly accommodated.

The control cabin will be positioned adjacent to the back wall of the site so that access to the cable trench on the adjacent wharf can be easily achieved for installing mains cable connection to the site network. All metal structures, cable tray supports and equipment in the control cabin will be connected solidly to a site copper earth mat. This earth mat will be solidly connected to the Marchwood earthing system. It is considered that a special lightning protection scheme is not required and all equipment is solidly bonded to earth.

8 SITE TRIALS, MAINTENANCE

After commissioning and acceptance by the EEC, the system will be operated for long periods in each of the four operating modes such that extensive data can be gathered and correlated to assess the potential of solar generation in the UK latitudes and to gain operating experience both in grid connected and standalone generating modes. During these trials, the system will be arranged to operate automatically and unattended but selected alarm facilities will be available at a manned control position in the CEGB Power Station complex so that system maloperation conditions can be quickly detected and the necessary arrangements made for servicing.

A routine maintenance procedure for the equipment will be prepared to include for annual cleaning of the array modules, regular topping up of the battery cells - the frequency will depend upon the operating mode - and limited maintenance within the control cabin, changing of PCU air filters,etc. The maintenance requirements are not onerous and will probably be undertaken by CEGB site staff.

9 PROGRAMME

As of early July 1983, the site construction phase is about to commence. It is anticipated that the system will be commissioned in October 1983 ready for acceptance by the Commission in November 1983.

ACKNOWLEDGEMENT

To Mr G. GILMORE, Digital Applications International Ltd. for contribution to "Monitoring" section of this paper.

LIST OF PARTICIPANTS

AUBERT, B.	SERI Renault Ingéniérie 2, avenue du Vieil Etang F - 78391 BOIS D'ARCY CEDEX Tel. (3) 043 8134 Telex 695 426 f
BUNNSDON, P.	BP Britannic House Moorgate UK - LONDON EC2Y 9BU Tel. 01/920 38 16 Telex
CORDES, V.	AEG - Telefunken Fachbereich Neue Technologien, Raumfahrt Industriestr. 29 D - 2000 WEDEL/Holstein Tel. 04103 - 702 - 833 Telex 2189520
DEMONGEOT, F.	PHOTOWATT 131, Route de l'Empereur F - 92300 RUEIL MALMAISON Tel. 1/708 05 05 Telex
DUMONT, P.	ACEC - Chevetogne Project ACEC 39.4 - URE/ENG P.O.Box 4 B - 6000 CHARLEROI Tel. 071/44 26 45 Telex
GILMORE, G.	Digital Applications International Ltd. Axtell House 24 Warwick Street UK - LONDON WIR 5RB Tel. 01/734 54 86 Telex
GOUEL, P.	PHOTOWATT International S.A. 131, route de l'Empereur F - 92500 RUEIL-MALMAISON Tel. 1/708 05 05 Telex 202084 f

GRASSI, G.

Commission of the European Communities
Directorate General Science, Research and Development
200, rue de la Loi - SDME 03/18
B - 1049 BRUSSELS
Tel. 235 68 01
Telex 21877 comeu b

GRUETER, J.W.

International Bureau
Nuclear Research Centre Jülich GmbH
P.O.Box 1913
D - 5170 JUELICH
Tel. 2461/61 30 10
Telex 833 556 30 kfa d

HELM, P.

WIP - München
Sylvensteinstrasse 2
D - 8000 MUENCHEN 70
Tel. 089/77 60 41
Telex 5212186

HENN, K.

KFA Jülich
Postfach 1913
D - 5170 JUELICH
Tel. 02461/61 48 80
Telex 833 556 30 kfa d

KOELEMEIJER, G.

HOLEC Projects
P.O.Box 258
NL - 7550 AG HENGELO
Tel. 074/45 16 96
Telex 44307 hml nl

KORUPP, K.-H.

AEG - Telefunken
Fachbereich Neue Technologien, Raumfahrt
Instustriestrasse 29
D - 2000 WEDEL/Holstein
Tel. 04103/702 846
Telex 2189520

KREBS, K.-H.

Commission of the European Communities
Joint Research Centre
Ispra Establishment
I - 21020 ISPRA(Varese)
Tel. 0039/332/78 99 02
Telex 380042 eur i

LOWALT, H.-J.

AEG - Telefunken
Fachbereich Neue Technologien, Raumfahrt
Industriestrasse 29
D - 2000 WEDEL/Holstein
Tel. 04103/702 786
Telex 2189520

MACOMBER, H.L.

MONEGON Ltd
4, Professional Drive, Suite 130
USA - GAITHERSBURG, Maryland 20760
Tel. 301-258-7540
Telex 710-828-9617

MERCIER, D.

Moteurs Leroy Somer
Usine des Agriers
Boulevard Marcellin Leroy
F - 16015 ANGOULEME CEDEX
Tel. 45/95 70 66
Telex 790 244 f

MERLIMA, S.

PRAGMA SpA
Via Po 25/A
I - ROME
Tel. 988 22 46/7/8 (NRTTONO) - 86 32 46 (Rome)
Telex

NICOLAY, D.

Commission of the European Communities
Directorate General Information Market and Innovation
Jean Monnet Building B5/005 - P.O.B. 1907
L - 2920 LUXEMBOURG
Tel. 43011 x 2946
Telex 3423/3446 comeur lu

PALZ, W.

Commission of the European Communities
Directorate General Science, Research and Development
Wetstraat 200
B - 1049 BRUSSELS
Tel. 735 00 40 - 735 80 40 - 736 60 00
Telex 21877 comeu b

PELINO, C.

Italenergie S.R.L.
Viale Repubblica 39
I - 67039 SULMONA
Tel. 0039 864/52 934
Telex

PREVI, A.

ENEL - Milano
Via A. Volta, 1
I - COLOGNO MONZESE (MI)
Tel. 02/8847 - 5255
Telex 310496 enel mi

RITTER, H.

SIEMENS AG - Dept. E 116
Werner-von-Siemensstrasse 50
D - 8520 ERLANGEN
Tel. 09131/72 37 73
Telex

SCHNELL, W.	Commission of the European Communities Directorate General Science, Research and Development 200, rue de la Loi, SDME 03/14 B - 1049 BRUSSELS Tel. 235 56 97 Telex 21877 comeu b
SCOTT, R.D.W.	LUCAS BP Solar Systems Ltd Windmill Road UK - HADDENHAM, Aylesbury, Bucks Tel. 0844/29 16 81 Telex 83165 lxshad
SELLèS, L.	SERI Renault Ingéniérie 2, av. du Vieil Etang B.P. 19 - Parc d'activités Sud F - 78391 BOIS D'ARCY CEDEX Tel. 33/3 043 81 34 Telex 695 426 serqy f
STRUB, A.	Commission of the European Communities Directorate General Science, Research and Development 200, rue de la Loi - SDME 03/25 B - 1049 BRUSSELS Tel. 235 46 83 Telex 21877 comeu b
TRANQUILLI, A.	PRAGMA SpA Via Po, 25 A I - ROME Tel. 86 32 46 - 86 34 27 Telex
TRäDER, K.	WIP - München Sylvensteinstrasse 2 D - 8000 MUENCHEN 70 Tel. 089/77 60 41 Telex 5212186
VAN GYSEL, M.	I.D.E. Parc Industriel B - 5430 ROCHEFORT Tel. 084/21 05 71 Telex 41969
VAN HAL, P.	GENSUN - C50/Hoboken E.N.I. Kontichse Steenweg 25 B - 2630 AARTSELAAR Tel. 03/887 40 81 Telex 31598

VIAUT, W.

Commission of the European Communities
Directorate General Energy
200, rue de la Loi
B - 1049 BRUSSELS
Tel. 235 39 70
Telex 21877 comeu b

VIEHWEG, J.

KFA - Jülich
Postfach 1913
D - 5170 JUELICH
Tel. 02461/ 61 48 76
Telex 833 556 30 kfa d

WELSCHEN, J.M.M.

Holecsol Systems
Lage Dijk 26
NL - HELMOND
Tel. 04920/48 886
Telex

WRIXON, G.

National Microelectronics Research Centre
University College
IRL - CORK
Tel. 021/26 871 x 2256
Telex 26050

ZOLLENKOPF, K.

Hamburgische Electricitätswerke
Ueberseering 12
D - 2000 HAMBURG 60
Tel. 040/636 31 53
Telex

GPSR Compliance

The European Union's (EU) General Product Safety Regulation (GPSR) is a set of rules that requires consumer products to be safe and our obligations to ensure this.

If you have any concerns about our products, you can contact us on ProductSafety@springernature.com

In case Publisher is established outside the EU, the EU authorized representative is:

Springer Nature Customer Service Center GmbH
Europaplatz 3
69115 Heidelberg, Germany

Batch number: 10607222

Printed by Printforce, the Netherlands